D0857698

Elements of the System Dynamics Method

MIT Press/Wright-Allen Series in System Dynamics

Jay W. Forrester, editor

Industrial Dynamics, Jay W. Forrester, 1961

Growth of New Product: Effects of Capacity-Acquisition Policies, Ole C. Nord, 1963

Resource Acquisition in Corporate Growth, David W. Packer, 1964

Principles of Systems,* Jay W. Forrester, 1968

Urban Dynamics, Jay W. Forrester, 1969

Dynamics of Commodity Production Cycles,* Dennis L. Meadows, 1970

World Dynamics,* Jay W. Forrester, 1971

The Life Cycle of Economic Development,* Nathan B. Forrester, 1973

Toward Global Equilibrium: Collected Papers,* Dennis L. Meadows and Donella H. Meadows, eds.

Study Notes in System Dynamics,* Michael R. Goodman, 1974

Readings in Urban Dynamics (vol. 1),* Nathaniel J. Mass, ed., 1974

Dynamics of Growth in a Finite World,* Dennis L. Meadows, William W. Behrens III, Donella H. Meadows, Roger F. Naill, Jørgen Randers, and Erich K. O. Zahn, 1974

Collected Papers of Jay W. Forrester,* Jay W. Forrester, 1975

Economic Cycles: An Analysis of Underlying Causes,* Nathaniel J. Mass, 1975

Readings in Urban Dynamics (vol. 2),* Walter W. Schroeder III, Robert E. Sweeney, and Louis Edward Alfeld, eds., 1975

Introduction to Urban Dynamics,* Louis Alfeld and Alan K. Graham, 1976

DYNAMO User's Manual (5th ed.), Alexander L. Pugh III, 1976

Elements of the System Dynamics Method, Jørgen Randers, ed., 1980

*Originally published by Wright-Allen Press and now distributed by The MIT Press

Elements of the System Dynamics Method

edited by Jørgen Randers

The MIT Press
Cambridge, Massachusetts, and London, England

This book was set in Dymo Times Roman by Allied Systems, Inc., and printed and bound by The Murray Printing Company in the United States of America.

Library of Congress Cataloging in Publication Data

International Conference on System Dynamics, Geilo,
Norway, 1976.
Elements of the system dynamics method.

(MIT Press/Wright-Allen series in system dynamics)
Includes bibliographies and index.
1. Social sciences—Mathematical models—Congresses.
2. System analysis—Congresses. I. Randers, Jørgen.
II. Title: System dynamics method.
H61.I583 1976 300'.1'51 79-20019
ISBN 0-262-18092-8

In memory of Gil Low

Contents

Part IV
Formulation

Part V
Testing

Part VI
Implementation

List of Contributors

David F. Andersen
Graduate School of Public Affairs
State University of New York at Albany
Albany, New York

James A. Bell
Department of Philosophy
University of South Florida
Tampa, Florida

James F. Bell
Retired Executive
Mobil International Corp.
Mobil Oil Corp.
New York, N.Y.

Alan K. Graham
Alfred P. Sloan School of Management
Massachusetts Institute of Technology
Cambridge
Massachusetts

Margaret S. Hamilton
Office of Resource Analysis
U.S. Geological Survey
Reston, Virginia

Gilbert W. Low
System Dynamics Group
Alfred P. Sloan School of Management
Massachusetts Institute of Technology
Cambridge
Massachusetts

Nathaniel J. Mass
Alfred P. Sloan School of Management
Massachusetts Institute of Technology
Cambridge
Massachusetts

Donella H. Meadows
System Dynamics Group
Dartmouth College
Hanover
New Hampshire

David W. Peterson
Pugh-Roberts Associates, Inc.
Cambridge
Massachusetts

Jørgen Randers
Resource Policy Group
Gaustadalléen 30
Oslo 3, Norway

Jennifer M. Robinson
System Dynamics Group
Thayer School of Engineering
Dartmouth College
Hanover
New Hampshire

Peter H. Senge
Alfred P. Sloan School of Management
Massachusetts Institute of Technology
Cambridge
Massachusetts

Lennart Stenberg
Resource Policy Group
Oslo 3, Norway

Carsten Tank-Nielsen
Resource Policy Group
Oslo 3, Norway

Henry Birdseye Weil
Pugh-Roberts Associates, Inc.
Cambridge
Massachusetts

Introduction

How are system dynamics models made? That is the question we try to answer in this volume.

We look at the question from the perspective of a practicing modeler. How does one choose a problem that will yield interesting results? What should be included in the model, and what can be left out? What amount of detail is desirable? How does one select parameter values? How does one know whether the model is "good"? And finally, what must be done to make the model interesting to other people? In short, this is a practical book attempting to give practical advice both to the novice and to the somewhat experienced system dynamicist.

The book also addresses modelers and policy analysts outside the system dynamics profession. For these groups, the volume serves to explain the rationale behind the untraditional (to use a pleasant word) behavior of the typical system dynamicist. In spite of its uncommonly soft and wide-ranging nature, the system dynamics school of modeling does have its own set of strict rules for what constitutes proper professional procedure. True, some rules deviate considerably and often contradict those widely used by other modeling schools. But this does not mean there are no modeling standards in system dynamics, as many critics seem to believe. The wealth of guidelines and norms presented in this volume should be sufficient proof.

In fact, there probably exists a more internally consistent system of guidelines and standards in system dynamics than in most modeling schools of comparable maturity. The unusually noxious criticism of the system dynamics school of modeling has forced upon the school a highly introspective attitude, with frequent debates of what one should choose as proper system dynamics practice.

This volume reaps the fruits of one such methodological debate. But since these debates certainly continue, the material in the following pages should not be viewed as final. It represents a state-of-the-art picture of system dynamics modeling. The main characteristics of the picture are likely to remain stable, however. And since this volume concentrates on fairly fundamental issues, it seems unlikely that its contents will be contradicted within the next decade.

Very little has been written about the actual process through which models are created and about the tricks that increase the chance of making a good model in a relatively short time. Aside from a lone doctorate dissertation,[1] this volume appears to be the first attempt at putting into print elements of the system dynamics method. Up until now, the novice has been left to develop his own working habits—often with no other guidance than the sequence of presentation in papers describing completed models. However, there is little semblance between a pedagogical ordering of stages in a presentation of a finished model and the recursive process of design, test, and dismissal in a productive modeling procedure.

It is probably fair to say that most of the professional debate around the system dynamics approach, as presented in the essays in this volume, stem from disagreements about the objectives sought in modeling studies. Since most mathematical models of social systems are constructed to generate high-precision forecasts, many critics seem to assume that *all* mathematical models are made for this purpose. When system dynamics models are made to satisfy other objectives, it should not come as a surprise that both the models and the modeling procedures are found wanting from the point of view of forecasting.

Two common objectives other than prediction are to increase understanding of some observed phenomenon (such as regular fluctuations in hog prices or accelerating stagflation) and to establish the general consequences of different options available at a decision point (such as the effects on profitability and employment of various policies intended to aid an ailing industry).

It is trite, but useful, to point out that a modeling procedure will differ according to the objective it is intended to satisfy. Any model has several characteristics; the model objective determines which of these will be emphasized. The following list presents some model characteristics that are commonly found desirable:

- *Insight generating capacity*. Does the model increase understanding of the modeled system? Does it improve the mental models of the model builders or the model clients? Does it produce surprising effects that are obvious after the fact?

• *Descriptive realism.* Do the model components and equations represent the real system in a form that corresponds closely to how persons experienced with the system perceive it? Does each parameter or element have a readily perceivable or conceivable real-world equivalent?
• *Mode reproduction ability.* Can the model produce important modes of dynamic behavior observable in the real system, under the same conditions that produce such modes in the real system?
• *Transparency.* Is the model easily understandable even by a non-professional audience? Does the model highlight the essential structure of the real system in an accessible way?
• *Relevance.* Does the model address problems viewed as important by experienced persons in the real system?
• *Ease of enrichment.* Can the model be altered to incorporate new findings or to test the effects of new policies not under consideration when the model was made? Can the model be adapted to represent systems related but not identical to the system originally represented? Can the model be updated without repeating all the work that went into its creation?
• *Fertility.* Does the model generate new ideas, new ways of looking at the problem, new experiments, or new policies that might have been overlooked in the absence of a model?
• *Formal correspondence with data.* Does the model incorporate real-world observations embodied in standard data sources, and can it reproduce under historical conditions a reasonable statistical fit to historically observed data?
• *Point predictive ability.* Can the model produce a precise prediction of a future event or of the future magnitude of important elements in the system?

The list is repeated in a diagram which also illustrates the different emphasis put on the various characteristics in two hypothetical modeling studies. The utility or quality of a given model depends on the extent to which it scores on those characteristics that a person judging it might view as important. Judges with different ultimate objectives obviously will give different marks to the same model.

In system dynamics much weight is typically placed on the upper entries in the list, whereas the lower ones are seen as less critical. This is rational when the ultimate objective is increased understanding, both of the past and of the likely consequences of future actions. The system dynamics approach is finely tuned to this objective and is therefore less useful if the goal is short-term, high-precision forecasting.

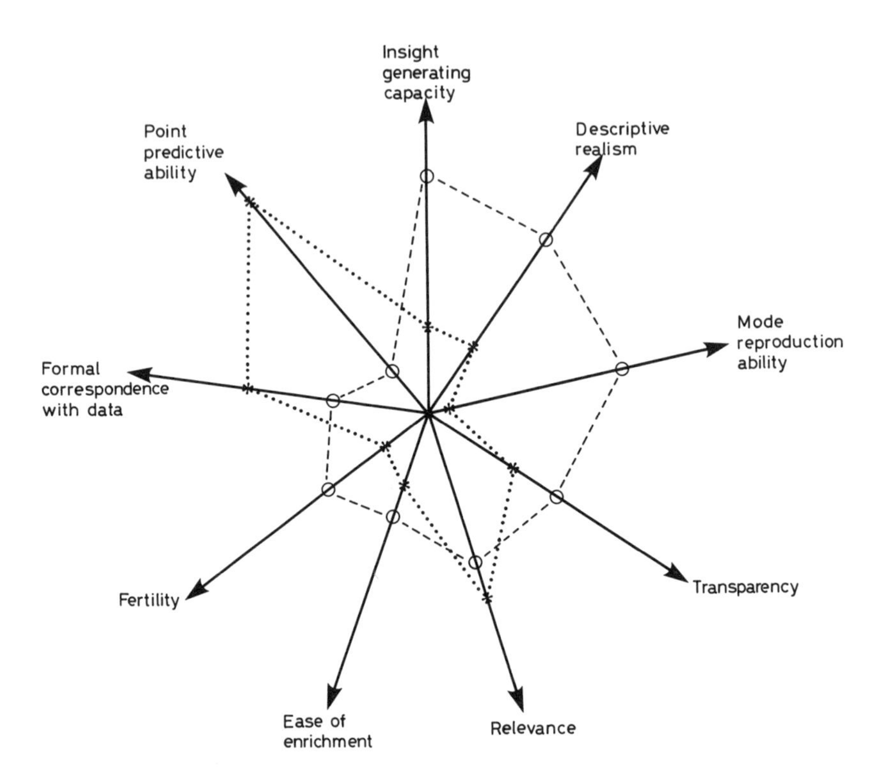

The star of model characteristics contrasting desirable modeling objectives for prediction (dotted line) and for increased understanding (dashed line)

The first group of papers in this volume, gathered in Part I, Paradigm, describes the objectives sought in system dynamics and the rationale behind the chosen objectives. Part II, Applied Principles, shows how application of the system dynamics paradigm actually result in different models and conclusions.

The remaining parts focus on the four stages in the process of model construction. Part III, Conceptualization, describes guidelines for the most difficult and least structured step in modeling, namely, that of establishing a well-structured mental understanding of the problem under study. Or in other words, to establish an effective mental model that also serves well as a basis for model development. Part IV, Formulation, gives practical advice on how understanding of the real world can be translated into model equations and parameter values. Part V, Testing, discusses various methods for evaluation of the resulting model: its selection of variables, the included relations, its sensitivity, and its overall behavior. Finally, Part VI, Implementation, gives educational glimpses of modeling histories (we could be tempted to say, "disasters") and presents hard won insights on how to proceed if the modeler not only wants to finish a modeling study before time, money, or interest runs out but also wants somebody else to be interested in and learn from the study.

This volume arose out of the 1976 International Conference on System Dynamics held at Geilo, Norway, August 8–15, 1976. Thirty-seven papers were presented at the conference, 30 were included in the proceedings,[2] and the present volume consists of a further subset of 14, largely rewritten, contributions. I would like to thank all participants in the Geilo conference for their contributions to *Elements of the System Dynamics Method*, either in writing, or through the week of discussions at Geilo, which certainly must have been one of the highlights in the history of system dynamics methodology.

Jørgen Randers

Notes

1. Jørgen Randers, 1973, *Conceptualizing Dynamic Models of Social Systems. Lessons from a Study of Social Change.* Ph.D. dissertation, MIT Sloan School of Management, Cambridge, Mass. (Available from the Resource Policy Group, Gaustadalléen 30, Oslo 3, Norway.)

2. Jørgen Randers and Leif Ervik, eds., 1976, *The System Dynamics Method*, Proceedings of the 1976 International Conference on System Dynamics, Geilo, Norway, August 8–15, 1976.

Published Works in System Dynamics

Alfeld, Louis Edward, and Alan K. Graham. 1976. *Introduction to Urban Dynamics.* Cambridge, Mass.: Wright-Allen Press, Inc.

Case, Fred E. 1974. *Real Estate Economics: A Systemic Introduction.* Los Angeles: California Association of Realtors, 505 Shatto Place.

Coyle, R. Geoff. 1976. *Management System Dynamics.* John Wiley and Sons.

Coyle, R. Geoff, and John A. Sharp. 1976. *System Dynamics—Problems, Cases, and Research.* John Wiley and Sons.

Forrester, Jay W. 1961. *Industrial Dynamics.* Cambridge, Mass.: MIT Press.

Forrester, Jay W. 1968. *Principles of Systems.* Cambridge, Mass.: Wright-Allen Press, Inc.

Forrester, Jay W. 1969. *Urban Dynamics.* Cambridge, Mass.: MIT Press.

Forrester, Jay W. 1971. *World Dynamics.* Cambridge, Mass.: Wright-Allen Press, Inc.

Forrester, Jay W. 1976. *Collected Papers of Jay W. Forrester.* Cambridge, Mass.: Wright-Allen Press, Inc.

Forrester, Nathan B. 1972. *The Life Cycle of Economic Development.* Cambridge, Mass.: Wright-Allen Press, Inc.

Goodman, Michael R. 1974. *Study Notes in System Dynamics.* Cambridge, Mass.: Wright-Allen Press, Inc.

Hamilton, H. R., S. E. Goldstone, J. W. Milliman, A. L. Pugh III, E. B. Roberts, and W. Zellner. 1969. *System Simulation for Regional Analysis: An Application to River Basin Planning.* Cambridge, Mass.: MIT Press.

Jarmain, W. Edwin, ed. 1963. *Problems in Industrial Dynamics.* Cambridge, Mass.: MIT Press.

Levin, Gilbert, Gary B. Hirsch, and Edward B. Roberts. 1975. *The Persistent Poppy: A Computer Aided Search for Heroin Policy.* Cambridge, Mass.: Ballinger Publishing Co.

Levin, Gilbert, Edward B. Roberts, Gary B. Hirsch, Deborah S. Klingler, Nancy Roberts, and Jack F. Wilder. 1976. *The Dynamics of Human Service Delivery.* Cambridge, Mass.: Ballinger Publishing Co.

Mass, Nathaniel J., ed. 1974. *Readings in Urban Dynamics: Volume 1.* Cambridge, Mass.: Wright-Allen Press, Inc.

Mass, Nathaniel J. 1976. *Economic Cycles: An Analysis of Underlying Causes.* Cambridge, Mass.: Wright-Allen Press, Inc.

Meadows, Dennis L. 1970. *Dynamics of Commodity Production Cycles.* Cambridge, Mass.: Wright-Allen Press, Inc.

Meadows, Dennis L., and Donella H. Meadows, eds. 1973. *Toward Global Equilibrium: Collected Papers.* Cambridge, Mass.: Wright-Allen Press, Inc.

Meadows, Dennis L., et al. 1974. *Dynamics of Growth in a Finite World.* Cambridge, Mass.: Wright-Allen Press, Inc.

Meadows, Donella H., Dennis L. Meadows, Jørgen Randers, and William W. Behrens III. 1972. *The Limits to Growth.* New York: Universe Books, A Potomac Associates Book.

Milling, Peter. 1974. *Der technische Fortschritt beim Produktionsprozess: Ein dynamisches Modell für innovative Industrie-unternehmen.* Wiesbaden: Dr. Th. Gabler Verlag.

Naill, Roger. 1977. *Managing the Energy Transition.* Cambridge, Mass.: Ballinger Publishing Co.

Nord, Ole C. 1963. *Growth of a New Product: Effects of Capacity-Acquisition Policies.* Cambridge, Mass.: MIT Press.

Packer, David W. 1964. *Resource Acquisition in Corporate Growth.* Cambridge, Mass.: MIT Press.

Pugh, Alexander L., III. 1973. *DYNAMO II User's Manual.* 4th ed. Cambridge, Mass.: MIT Press.

Randers, Jørgen. 1973. *Conceptualizing Dynamic Models of Social Systems: Lessons from a Study of Social Change.* Ph.D. dissertation, Alfred Sloan School of Management, MIT, Cambridge, Mass.

Roberts, Edward B. 1964. *The Dynamics of Research and Development.* New York: Harper and Row.

Roberts, Edward B., ed. 1978. *Managerial Applications of System Dynamics.* Cambridge, Mass.: MIT Press/Wright-Allen Press, Inc.

Schroeder, Walter W. III. 1974. *Urban Dynamics in Lowell.* Cambridge, Mass.: System Dynamics Group, E40–253, MIT.

Schroeder, Walter W. III, Robert Sweeney, and Louis E. Alfeld, eds. 1975. *Readings in Urban Dynamics: Volume 2.* Cambridge, Mass.: Wright-Allen Press, Inc.

Weymar, F. Helmut. 1968. *The Dynamics of the World Cocoa Market.* Cambridge, Mass.: MIT Press.

Zahn, Erich. 1971. *Das Wachstum industrieller Unternehmen—ein Versuch seiner Ecklarung mit Hilfe eines komplexen, dynamischen Modells.* Wiesbaden: Dr. Th. Gabler Verlag.

Part I
Paradigm

System Dynamics and Scientific Method

1

James A. Bell
and
James F. Bell

1.1 Introduction

Recognition that science is not absolute truth raises a question: Then what is it? The practice of system dynamics and its alternatives is deeply affected by three views that have emerged as answers. One view maintains that scientific knowledge is a tool—an instrument—that consists of data correlations. It is advanced by correlating more and more data and/or by finding correlations with a closer statistical fit to the data. This view, instrumentalism, underlies much work in the social sciences, including econometrics. According to another view, scientific knowledge consists of groups of ideas—paradigms—that explain phenomena. It advances by forcing the ideas on data until serious misfits require adoption of another paradigm. Paradigmism has intrigued numerous social scientists, including system dynamicists. The third view holds that scientific knowledge consists of conjectures that are refutable—vulnerable to empirical error—and that its principal vehicle of advance is to adjust or change mistaken conjectures to overcome refutations. Refutationism is the view used in good system dynamics practice. Despite its long and successful tradition in the physical sciences, it is not widely embraced by social scientists.

Each view of knowledge functions as method: techniques to improve theories and confidence in those theories are inextricably linked to them. The methods of these different views are inconsistent, however, because they render varying guidelines for advancing knowledge and incompatible criteria for judging knowledge "scientific." The broad purpose of this chapter is to unravel the controversy that has ensued.

The advantages and disadvantages of each view for advancing knowledge and for confidence in that knowledge will be the focus here. An assessment of the pros and cons should be the primary factor in deciding which view of knowledge to employ for empirical work. Unfortunately, historical reasons often predominate. There will be a journey through the history of the different views prior to assessing the pros and cons of each. It should help explain the fervor of allegiance that can grip adherents of a view as well as provide a context for presentation of those views. The final section will explore refutationist method further in order to throw more light on pivotal ideas and attitudes beneficial for good practice of system dynamics.

1.2 Functions of Method

Method can function in four identifiable ways: to help generate knowledge, legitimate ideas, render other ideas suspect, and propagate ideas to others. Each of the four functions—generation, legitimation, suspicion, and propagation—are illustrated by contrasting the method implied by two well-known views of knowledge in the history of philosophy: empiricism and rationalism.

The first function of method is to provide a formula for generating knowledge. An empiricist, as a first step in gaining knowledge, collects facts through observation. Only after exhaustive collection of empirical data will he induce ideas from it, and even then he will be skeptical of his thought processes. A rationalist, on the other hand, believes that the first stage in generating knowledge is to think. Knowledge is the jewel of precise and clear thoughts. Skeptical of observational data, he uses it reluctantly and only then to clarify his ideas. If there is conflict between data and ideas, he tends to trust the latter.

The second function of method is to legitimate ideas. An empiricist claims legitimacy for his ideas because they are based on data, and thus labels them "scientific." A rationalist legitimates his ideas not on the basis of empirical research but on the precision of his thinking and the intrinsic clarity of his ideas. He also labels his ideas "scientific." Each accords scientific status by a different criterion.

A corollary of the legitimating function is the third function of method: suspicion of ideas generated, or legitimated, by other methods. Empiricists quickly attack ideas not thoroughly grounded in observation. Thinking is debilitating and extended contemplation is particularly dangerous. Rationalists, on the other hand, are not impressed with empiricist claims to scientific status. Facts are misleading, and ideas rooted in them are just as suspect.

Dialogues between empiricists and rationalists are sometimes barren. A claim to science by one is often an invitation to suspicion for the other. One classic debate between empiricist followers of Newton and rationalist followers of Descartes provides an example. Newtonians believed there was a measurable force between masses, gravitational force, and viewed the force as a fact. Whether the force acted through empty space or required an ether for its propagation was debated, but the fact that gravitation existed was beyond dispute. Cartesians, on the other hand, deduced that there could be no gravitational force from their theory that all motion results from pushes. The "fact" of gravitation, which merited scientific status for the Newtonians, was rendered suspect by the deductions of the Cartesians.

Propagation is the fourth function of method. The empiricist mandate for propagating ideas is to encourage observation of the facts. Empiricist indoctrination is a factual baptism: an absolution of preconceived notions and a new start based on facts. Books, reports, and other presentations begin with research findings. Conjectures and speculations are confined to a secondary role and appear, if at all, at the end. Even there, they are introduced by cautionary warnings and even apologies.

Rationalists propagate their ideas by invoking one's abilities to think clearly and draw conclusions validly. Like empiricists they ask that preconceived ideas and prejudices be discarded. Their appeal, however, is to one's intuition and common sense, not to the facts. Most deductivist books and reports commence with axioms followed by deductions. Sometimes there is an introductory attack on other ideas to show them unclear, inconsistent, or in conflict with intuition or common sense.

Each view of knowledge discussed in this chapter—instrumentalism, paradigmism, and refutationism—implies its own method. Each method, however, functions in the same four ways.

Generation and legitimation are pre-eminent functions. Arguments in support of, or against, a view of knowledge usually turn upon how well its method is performing these functions. Suspicion and propagation are derivative. Nevertheless, at stake is the ability of an intellectual community to accept and learn from external criticism, and its effectiveness in teaching and winning acceptance for its ideas. A history encompassing the third and fourth functions would be replete with secret intrigues, heretical expulsions, and other scandalous episodes. Such tales are spicy and absorbing, but this chapter will confine itself to the principal functions of generation and legitimation.

1.3 Historical Perspective

Nothing is stronger than habit.—Ovid (43 B.C.–A.D.18)

There is a tale about a devout tribe of natives isolated by the jungles of Brazil. Torrential rains had flooded their land, devastating their homes, crops, and livestock. Starving and without shelter, they reverently climbed a sacred hill to implore their gods for mercy. During the ceremony, when all eyes were searching for an omen, a relief plane came up from the southern horizon. They watched in awe as the soundless, tiny speck grew to a roaring giant above their heads. As the plane dropped packages of food and tools, the natives bowed in subservience to their savior.

With solemn faith the tribe erected a crude replica of their new, merciful god, and with the tenacity of unquestioning believers, repeated the ceremony to their totem day after day, month after month—always expecting him to reappear with another cargo.

It would not be fair to assert that method is observed as diligently as the rituals of these natives, nor that a scientist's faith in a delivery of knowledge is as strong as the natives' faith in a delivery of cargo. Scientists often cling so firmly to method, however, that it is not totally unfair to interpret their actions as those of a "cargo cult": a phrase coined by anthropologists to denote a group which believes proper rituals will yield an expected result. An understanding of the historical roots of different methods will make the tenacity less surprising. Adherents believe that their method best explains developments in science and hence believe that it must be valid.

Francis Bacon's inductive method, we shall see, was thought to have delivered absolute truth—truth that was beyond all doubt and which would be correct for eternity. The overwhelming success of Newton's physics gave credence to that belief until the Newtonian program finally cracked definitively at the beginning of the twentieth century. There had been skeptics of Baconian induction before that time; Hume's attack laid the foundation for instrumentalism, and Whewell's criticism led to a version of paradigmism. It was the establishment of relativity theory and quantum principles in place of Newton's physics, however, that fertilized the grounds for acceptance of the new methods.

Francis Bacon's Revolution in Scientific Method

There are and can be only two ways of searching into and discovering truth. The one flies from the senses and particulars to the most general axioms . . . this way is now in fashion. The other derives axioms from the senses and particulars, rising by a gradual and unbroken ascent, so that it arrives at the most general axioms last of all. This is the true way, but as yet untested.—Francis Bacon, *Novum Organum* (1620)

Francis Bacon (1561–1626), Lord Chancellor under James I of England, distinguished himself in law, literature, politics, and philosophy. Perhaps his greatest contribution, however, was a new theory of how to seek and advance knowledge.

The centers of learning in England during Bacon's era were universities, the strongholds of Catholic thought. Even though the Anglican Church had already been formed under the tumultuous reign of Henry VIII, both Cambridge and Oxford still thrived as Catholic centers. At a time when it was fashionable to blame the Catholic Church for nearly all that seemed authoritarian, Bacon condemned Catholic theology in general and its Aristotelean roots in particular for stifling the growth of knowledge.

In such works as *Advancement of Learning* (1605) and *Novum Organum* (1620) Bacon argued that natural philosophy—science—had progressed little since ancient times. He saw no merit in speculative philosophy, contending that in some respects modern thinkers knew less than the Greeks. Bacon did admire the revolutionary discoveries of Copernicus and Galileo, marveled at the explorations of Marco Polo and Magellan, and appreciated such inventions as the printing press and gun powder.

Contrasting the lack of progress by speculative thinkers to the remarkable gains of others, Bacon posed himself a question: What demarcates speculative thinking from progressive thinking? His answer can be summarized in two parts.

First, lack of progress in the Catholic tradition was due to speculation about essences. Aristoteleans believed that the universe consists of essences. Bacon contended that statements about them reflect subjective belief and not objective reality. Aristotelean science amounted to idle conjectures about the definitions of concepts as a consequence, and Aristotelean ideas pertained to the actual world only by accident. Progress could not be made by speculation.

Second, progress necessitated observation of material facts. The facts, unlike Aristotelean intuitions, would have to be located in the world outside ourselves. In sum, progress would be made by avoiding speculation and observing facts.

There is a place for thinking in Baconian method. After the fact-gathering stage of science, there was to be idea-extracting stage. This is induction: extrapolating general ideas from specific facts.

Bacon's method provides a legitimation criterion for deciding which ideas are scientific and which are not. Since all scientific ideas must be inductions from facts, all scientific ideas must be reducible to facts. Ideas not reducible to facts are not legitimate. Since Bacon's time, methods that assume that general ideas are generated from facts and/or are reducible to facts have been called "inductivist" methods. As shall be seen, the instrumentalist view of knowledge implies a version of inductivist method.

A crucial element in Bacon's view of knowledge is absolute truth. His inductive method was a tool to deliver the cargo, a means of guaranteeing that it would be obtained. He reasoned that if facts in the real world are beyond question, and scientific ideas are induced from facts, it follows that scientific ideas must also be beyond question—that they are absolute truth.

Despite advantages over Aristotelean disputation, inductivism was criticized and its shortcomings exposed. The attacks were crucial for the development of instrumentalism and paradigmism, but before investigating those attacks we should understand why inductivism became so highly regarded and widely accepted.

The Newtonian Fortress Protects Inductivism

A coincidence of circumstances surrounding and including Isaac Newton's physics allowed inductivism to become the predominate method by the late seventeenth century and sustained it through the nineteenth century. The sweeping success of Newton's dynamics and celestial mechanics, the adoption of inductivism by the Royal Society of London, and the anti-Catholic mood in England all conspired to establish Baconian method as the deliverer of absolute truth.

The story is typical. A method believed to have produced successful research is widely accepted. Whether or not it actually produced the research program is seldom asked. Even valid criticism is little noted until the research program has run its course and/or is replaced by another research program.

Sixty-one years after Bacon died, Isaac Newton published his *Mathematical Principles of Natural Philosophy* (1687). It outlined a system of dynamics and celestial mechanics that was corroborated on all fronts. Almost everyone believed that Newton's physics was absolute truth. Further, Newton's success was credited to *his adherence to inductive method.* The promised cargo—absolute truth—seemed to have been delivered. What more could be asked of method?

Baconian method had received a significant endorsement even before publication of Newton's *Mathematical Principles*: the Royal Society of London had adopted it as the proper and official formula for the advancement of knowledge. The Royal Society, one of the first scientific institutes, had been founded independent of university influence. An institution free of Catholic domination was believed crucial for the advancement of knowledge. As a fledgling group organized by such men as Robert Boyle, it passed through infancy in the unsettled times of the Civil War, the Protectorate under the Cromwells, and the Restoration under Charles II to become one of the most prestigious scientific institutions in the world.

By the eighteenth century, Isaac Newton had become the most famous and revered member of the Royal Society. Since the Royal Society had adopted Baconian method as its official formula for advancing knowledge, Newton's work appeared to be a product of induction. Despite Newton's known reservations about the viability of Baconian induction, it was commonly believed that he had used it.[1] Even the Royal Society, basking in the fabulous success and acclaim shining on Newton, did little to discourage the misconception.

There were critics who published their views. Two of the most profound critics of Baconian induction during the era of Newtonian successes were David Hume (1711–1760) and William Whewell (1794–1866). Hume's attack provided the arguments which led to instrumentalism. Whewell's historical psychological analyses, which were similar to those of Thomas Kuhn over one century later, are behind another alternative: paradigmism.

Hume Attacks with Logic

Never literary attempt was more unfortunate than my *Treatise of Human Nature.* It fell dead-born from the press.—David Hume, *My Own Life* (1777)

Initially David Hume, like nearly all his contemporaries, assumed that Newton's dynamics and celestial mechanics were absolute truth. He also assumed that inductive method had delivered the dynamics and celestial mechanics. After comparing these two assumptions, however, he found them inconsistent. Intellectuals gave Hume's *A Treatise of Human Nature* a very cold reception when it was published in 1739. His attack, after all, threw doubt on two cherished beliefs: the absolute truth of Newtonian physics and the validity of inductive method.

Hume argued that Newtonian physics presupposes a universal law of causality: for every event there is a cause. But, his reasoning continued, no one can be sure that in the future the same causes will lead to the same effects, or "facts." Hence future "facts" can only be inferred with a degree of probability. Hume concluded that (1) either Newtonian physics, if it originated in and is legitimated by inductive method, cannot be absolute truth; it can be highly probable at best; (2) or inductive method is not entirely valid; (3) or both (1) and (2).

Hume did make a hesitant choice between the two pillar beliefs. He put his faith in inductive method and, at the same time, declared Newtonian physics highly probable.

The view of knowledge attached to Baconian induction had changed forever after Hume's analysis. The assumption that knowledge is absolute truth could not be supported. Hume's attack led to a fallback position of inductivism: facts can generate ideas that are probable but not certain; legitimate ideas are reducible—within a reasonable degree of probability—to the facts. It is but a short step to the view that science is a statistical correlation of data. Baconian induction had given birth to instrumentalism. Hume was the midwife.

Whewell Analyzes Inductivism

The examination of the steps by which our ancestors acquired our intellectual state, may make us acquainted with our expectations as well as our possessions, may not only remind us of what we have, but may teach us how to improve and increase our store.—William Whewell, *History of the Inductive Sciences* (1837)

Nearly one century after Hume's attack, William Whewell put forth historical and psychological arguments against Baconian induction. Like Hume's work, Whewell's major essays, *History of the Inductive Sciences* (1837) and *Philosophy of the Inductive Sciences* (1840), set forth unpopular theses.

Whewell's studies in the history of science and psychology of discovery indicated that the processes of advancement did not resemble the research and induction stages prescribed by Bacon. Scientists, he found, made bold guesses and then tested them against facts. Most of the guesses turned out to be mistaken, but a few were correct. Even correct ideas, however, could not be proclaimed the absolute truth; they very well might be found mistaken by later tests. Whewell argued that freedom in thinking and a vivid imagination were important elements in the guessing and the testing.

Although facts seemed unimportant for generating knowledge, Whewell did maintain that they are significant for legitimating knowledge. When an idea had helped a scientist understand phenomena, he would then deduce the pertinent facts to confirm the idea. If the facts were there as expected, the idea would be legitimate. Unlike inductivism, however, legitimate ideas are not totally reducible to known facts. According to Whewell, ideas grouped into an intelligible whole usually entail facts which have not been foreseen.

Whewell's challenge to Baconian induction was clear: inductive method could neither adequately explain the generation nor the legitimation of knowledge. Whewell's view of knowledge is similar to one interpretation of Thomas Kuhn's paradigmism.

Inductivism Repels the Attacks

Despite the attacks by Hume and Whewell, induction did not surrender. The unprecedented success of the Newtonian research program and the belief it was produced by inductive method made a strong fortress. The only visible alteration was the fallback to probabilistic induction. While it was admitted that facts could not guarantee truth, but only likelihood, other basic ingredients of induction were maintained: ideas are generated by induction from facts and ideas are legitimated by a probabilistic reduction to the facts. Whewell had planted seeds for paradigmism, but those seeds remained in germination for over a century. Thomas Kuhn finally sprinkled water on them in the early 1960s.

In any case, the Newtonian bastion stoutly defended the harrassed induction. At the turn of the twentieth century, when Newtonian physics itself began to falter and was finally trampled by the new theories (in relativity and quantum phenomena), Baconian induction was sacked and destroyed.

The Instrumentalist View of Knowledge

Nothing at bottom is real—except humanity.—Auguste Comte, *Systeme de Politique Positive* (1851–1854)

Classical probabilists do not maintain that knowledge is absolute truth, but they do believe that probabilistic statements are about "facts" and that those facts are connected to the actual world. The assumption that facts connect us with the actual world was challenged, however, during the high tide of quantum mechanics. When Werner Heisenberg formulated his famous uncertainty principle—roughly, the uncertainty of the position of an electron is proportional to the certainty of its momentum, and vice versa—a totally new impasse arrived in the history of science. It seemed impossible that more precise knowledge of the electron could be gained because the measuring waves altered its position and/or momentum. Even though a statistical description of the actual world seemed possible, knowledge beyond that description seemed impossible. This impasse fit well with the positivistic interpretation of science; that is, scientific statements are merely instruments that should make no claims about a real world behind observable phenomena.

Instrumentalism thus became very fashionable. According to the instrumentalist view, knowledge consists of data correlated together. It is advanced by gathering more and more data and/or by finding correlations with a closer statistical fit to the data. Causal explanations of the data are not endorsed as part of science, although they might be considered as tools for generating a data base.

The instrumentalist view of knowledge was not new to the intellectual world. Even August Comte (1798–1857), who coined the word "positivism," was not the first to espouse instrumentalism. Long before Comte, in the sixteenth century, Andreas Osiander, who wrote a preface to Copernicus' *De revolutionibus*, claimed of Copernicus' heliocentric view of the universe that there is no need for Copernicus' hypotheses to be true, or even resemble truth; it was sufficient that they should produce calculations which agree with observations. Not until the beginning of the twentieth century, however, did instrumentalism take a firm hold on the scientific community.

The example that follows illustrates use of instrumentalist method. Those familiar with such fields as econometrics or classical behaviorism in psychology will probably notice the similarities, as would those familiar with the use of instrumentalist method in any other field.

Physical chemistry since World War I and chemical engineering since World War II have had dominant instrumentalist tendencies. In both fields there has been an impetus to generate equations for incorporating data with little or no conjecture about explanations of the equations.

Such equations are sometimes called "empirical equations." Data that might be inconsistent with the equations are generally handled in one of two ways: the equations are altered so that the data are incorporated, or some "boundary conditions" are imposed, that in essence, simply proclaim the data not pertinent to an equation.

The instrumentalist view of knowledge popular in physical chemistry and chemical engineering has made both fields fertile for purely mathematical manipulation, with results that have not been entirely beneficial. Since teaching methods usually reflect beliefs about scientific method, students are often encouraged by professors and textbooks to plug parameters into empirical equations. While this technique seems practical for routine work in both fields, it can impose limitations on the student, practicing engineer, or chemist when faced with an anomaly, or a totally new problem. Without adequate training in generating formulas, it is difficult to analyze the factors that result in anomaly or the factors that could lead to an entirely new formula. There are also reasons why instrumentalist method can seriously impede theoretical progress. A discussion of those reasons appears in section 2.4.

Ironically, instrumentalism is not a vogue amongst subatomic physicists as once was the case. The resistance of such physicists as Einstein, as well as its barrenness for theoretical progress, have perhaps been responsible. The cults endorsing instrumentalism are far from dead, however, as anyone familiar with modern social science can testify.

Paradigmism Reinvented in Thomas Kuhn's View of Knowledge

System dynamicists and other social scientists have been interested in Thomas Kuhn's view of knowledge. The adoption of his most crucial word—paradigm—into the system dynamics lexicon is indicative of the significance of his influences. It is important to be familiar with two interpretations of Kuhn's ideas, one of which is a vague version of the refutationist view discussed in the next section and the other of which offers only questionable method for science.

The following theses of Whewell will provide reference for this section:

1. Science grows by imaginative new ideas which are then used to search out facts.
2. Facts are only seen in light of these ideas; research is directed to uncover the facts.
3. Scientists try to group ideas into units.

4. There is a strong tendency to force the facts to fit one's ideas.
5. Approximate comparisons between competing ideas can be made by measuring their success in explaining facts; then a comparison can lead to the separation of mistaken ideas from other ideas.

Thomas Kuhn's *The Structure of Scientific Revolutions* (second edition, 1962) begins with a criticism of inductivist interpretations of the history of science and suggests that Kuhn's interpretations be considered as a replacement. Kuhn's views parallel the first four of Whewell's theses. The fifth Whewellian thesis, however, sometimes surfaces in Kuhn's work but at other times is deliberately drowned. The spasmodic affirmation and denial of the fifth thesis has led to two different interpretations. Let us see the details.

A given view of the world along with method, logic, and other background assumptions seem to constitute what Kuhn calls a "paradigm." Paradigms provide the basis for a research program, and "normal science" is devoted to working out that research program. Paradigms, and the normal science extending from them, operate according to the first four theses of Whewell. The fifth thesis of Whewell, however, Kuhn sometimes endorses:

1. A paradigm is thrown into doubt when an accumulating number of facts do not justify it.
2. A new paradigm must explain crucial facts that are anomalies in the old paradigm.
3. A new paradigm explains facts about which its predecessor implies nothing.

And he sometimes denies:

1. No paradigm can be reduced to another—mathematical formulations adhering to paradigms cannot even be reduced to others.
2. Adoption of one paradigm over another is ultimately a leap of faith.
3. Acceptance of a new paradigm is an emotional experience, not an intellectual transition.
4. Paradigms spread, and are accepted, because of sociological reasons—the fame of its inventor or supporters, the popularity of a textbook espousing it, and so forth.

In short, there are two conflicting interpretations: paradigms can be compared and selected by intellectual criteria, and they cannot. The pros and cons of both interpretations will be discussed in section 1.4.

Refutationism: Knowledge as Conjectures and Refutations

The Gods did not reveal, from the beginning, all things to us; but in the course of time, through seeking, men find that which is better.

But as for certain truth, no man has known it, nor will he know it; neither of the gods, nor yet of all the things of which I speak. And even if by chance he were to utter the final truth, he would himself not know it; for all is but a woven web of guesses.—Xenophanes, *Verses* (570–475 B.C., from Popper, 1963, p. 26).

Refutationism is another view of knowledge put forth after the fall of Newton's program. It also assumes that scientific knowledge is not absolute truth, but it is centered upon the thesis that scientists can find empirical error—refutations—and can use error as a springboard to improve their theories. Since exposing theories to potential error is so important, instrumentalism is criticized because of the reduced exposure to error—by placing boundaries on the data pertinent to theories, for example, or by accepting improbable but possible data in a statistical correlation—and the second interpretation of Kuhn's paradigmism is criticized because of the dogmatic tendency to cling to ideas in the face of empirical error rather than giving up ideas that have been refuted.

Like adherents of other views, refutationists believe that their method best explains developments in science even though it was not systematically formulated until Karl R. Popper published *The Logic of Scientific Discovery* in the early 1930s (original in German). Refutationism has been expanded and improved over the past five decades by Popper and his followers.

Refutationism is a clearer and more sophisticated version of the first interpretation of Kuhn's paradigmism. The overview here will serve for the discussion of pros and cons in the section 1.4. Since refutationist method is used in good system dynamics practice, a more detailed presentation will be given in section 1.5. In a nutshell, the refutationist view is as follows: knowledge consists of conjectures from which expectations can be deduced for empirical testing. Although scientists wishing to establish a theory hope the tests will be passed, a crucial quality of scientific conjectures is that they be refutable, or vulnerable to empirical error. If the conjectures pass empirical tests, they are corroborated; if not, they are falsified. In either case progress is made. Corroborated theories have made contact with reality because they have survived risky tests against empirical data; they have been shown empirically correct rather than incorrect at those points. Falsified theories are important because the points at which they have failed tests provide desiderata for improving theories. Improved theories are corroborated at the points where predecessors failed. Improved theories explain reality better than the predecessors because they have passed more risky tests against empirical data.

1.4 Pros and Cons of Instrumentalism, Paradigmism, Refutationism

That different views of knowledge claim to account for developments in the history of science better than others may explain the faith in them; but it does not leave any clear criterion for deciding which is preferable. In this section then further advantages and disadvantages of each view will be considered. The discussion will not be so abstract as to be useless because practice of econometrics is largely devoted to instrumentalist method and good practice of system dynamics employs refutationist method. The interpretation of Kuhn's paradigmism that is close to refutationism can be appropriately discussed in the section on refutationism. The other—or dogmatic—interpretation of Kuhn's method will be briefly discussed in the section on paradigmism.

Desiderata for Weighing Views of Knowledge

What criteria ought be used to decide the relative merits of views of knowledge? The following two seem appropriate:

1. *Problem-solving power.* Consideration should be given to the *types* of problem for which a view of knowledge can generate fruitful solutions. The *importance* of those problems should be weighed, despite the value judgments that will be inevitable. The *range* of problems for which a view of knowledge is useful should also be considered.

Since chapter 2 (by Meadows) is largely devoted to comparing the problem-solving qualities of system dynamics and its alternatives, this section will focus on the value of different views for theoretical progress: for increasing our understanding of the empirical world. It must be remembered, however, that theories which better explain the empirical world are often more powerful theories for solving problems rooted in the empirical world. Method which is better for theoretical progress is also often better for solving problems.

2. *Theoretical progress.* A more subtle factor when weighing views of knowledge is the impetus and/or constraint entailed for increasing our

understanding of the empirical world. The method implied by all views of knowledge will render formulas for generating, and criteria for legitimating, knowledge; what is desired is method that will generate and legitimate ideas likely to lead to the following: (a) new, and often unforeseen, insights, (b) theories that explain a great variety of phenomena and (c) theories that can be confidently judged as improvements over preceding theories.

There are other factors, of course. In addition to the role played by views of knowledge in suspicion of other ideas and propagation of endorsed ideas, they can be influential in the formation of social-political structures and personal attitudes. The assumption that knowledge is absolute truth, for example, might be detrimental for a liberal democracy. Leaders might be encouraged to think they know what is best for everyone (the truth), and citizens might not be inclined to execute their important role as critics of public policy. Such factors are regrettably beyond the scope of this chapter.

Instrumentalism

Perhaps the greatest advantage of instrumentalism for theoretical progress is its encouragement to search for correlations *without* being constrained to offer explanations for the correlations that were unforeseen. Looked at from another perspective: if one is obliged to explain data, then data that do not fit a model might be avoided. For example, an econometrician attempting to find correlations that constitute indicators of a future slowdown in capital investment could make correlation runs with all sorts of data, and some correlation(s) might become evident that had not even been considered. A noninstrumentalist, on the other hand, might already be committed to certain types of correlations that he infers from prior assumptions, such as the growth or contraction of money supply. The prior commitment might constrain the search for correlations that are not explained by money supply. New insights might be lost.

There is a significant impediment to theoretical progress for those using instrumentalism, however. *The instrumentalist view greatly lessens vulnerability to error.* The argument for this claim is complicated, but an intuitive version will be given here. The ramifications will be traced immediately thereafter.

Instrumentalist correlations are less vulnerable to error because, by making no commitment to an explanatory model of the correlations, there is no deduction of possible data that, if found, could corroborate or falsify the model. Data that might not correlate can be excluded by boundary conditions, "fudged" into the equation by mathematical manipulation, or explained as one of those improbable—but possible—exceptions.

There are a number of unfortunate consequences. First, models that run little empirical risk do not merit much confidence if they pass empirical tests. Potential contact with the empirical world is thus lessened. Second, because correlations are made from a given data base, the potential for correlations to explain a great variety of phenomena is limited. The potential to lead to new insights is limited for the same reason. Finally, since it is difficult to know where a model is in error, it is difficult to know if another model is an improvement. Unable to pinpoint empirical error, instrumentalists are forced to use agreed-upon conventions for rejecting models and accepting replacements, such as statistical rules. Compliance with a statistical rule does not assure that a replacement model has survived risky empirical tests where a predecessor has failed, even if it does assure a closer fit to the data to which it is designed to conform.

Paradigmism

The dogmatic interpretation of Kuhn outlined in section 1.3 implied that there is no intellectual criterion to weigh paradigms. Changes in paradigms are the result of sociological forces rather than rational comparison.

One psychological advantage is that practitioners of a new field can feel free to pursue their research even though it may rest upon assumptions and utilize techniques that are unacceptable in more established alternatives. Since paradigms cannot be rationally compared, criticisms across paradigms can be comfortably ignored as misguided or unimportant criticisms.

Another advantage is that practitioners are encouraged to pursue ideas until nearly every type of possible phenomena has been explored. In other words, the fruits of a research program would most likely be exhausted before a paradigm is rejected. The drawback, however, is that the paradigm might be kept long after its research program has been exhausted.

A disadvantage of the dogmatic interpretation is the lack of a clear criterion for deciding which amongst a number of paradigms is preferable. If paradigms cannot be intellectually compared, there can be no rational criterion for preferring one over others.

Other advantages and disadvantages for theoretical progress—for leading to new insights and explaining a variety of phenomena—are the same as for the other interpretation of Kuhn's ideas.

Refutationism

The first interpretation of Kuhn's ideas entailed that competing paradigms can be rationally compared: the preferred paradigm explains anomalies in another paradigm. Refutationism provides a more precise criterion: an improved theory will be corroborated where its predecessor has been refuted. In other words, an improved theory will make contact with the empirical world at points where its predecessor failed. This criterion presupposes that theories be vulnerable to empirical error, and the more refutable the better. That quality is important not only for providing a criterion to distinguish improved theories from those they supercede but also for the other ways in which refutationist method enhances theoretical progress. Let us explain.

Refutationists conject causal models. Causal explanations are important because they always contain universal statements, and universal statements are normally very vulnerable to empirical error. For example, the causal explanation "heat causes metal to oxidize (in the presence of oxygen)" contains a universal statement: "whenever heat is applied to metal, the metal will oxidize." Any instance of a metal being heated but not oxidized would refute the explanation. Causal explanations consisting of two or more universal statements become even more vulnerable to error because they expose more points for testing. For example, the Newtonian causal explanation for projectile motion (in a vacuum) involves two universal statements: (1) all masses attract each other (the gravitational law) and (2) all bodies in motion will stay in that motion until acted upon by an external force (the inertial principle). The causal explanation is refutable not only by all types of projectile motion but also indirectly by all other types of motion involving one or more of the two universal statements: planetary motion, pendulum motion, the motion of free-falling bodies, the motion of bodies on an incline plane, and so forth.

The more points for testing and the greater the variety of test points, the more potential a causal explanation has to explain a variety of phenomena. A causal explanation will explain a variety of phenomena if it is subsequently tested and corroborated at those points. Furthermore, the more points for testing the greater the potential for new insights and unforeseen connections. In the preceding example Newton's causal explanation with multiple universal statements explained a great variety of phenomena—all types of motion are explained by the same principles —and led to new and unforeseen insights, such as the relationship between projectile motion and pendulum motion.

In sum, the refutationist method seems best for theoretical progress. The concentration on causal explanations enables new insights to be found, a variety of phenomena to be explained, and provides a clear criterion for judging theory improvement. Neither instrumentalism nor dogmatic paradigmism does the same.

1.5 Conclusions

It is evident from the preceding discussion that application of refutationist method requires search for causal explanations. This idea and some of its implications will be pursued further in this concluding section.

Causal models are important because they are refutable. In addition, practitioners should *insist upon causal explanations of errors in models*. A causal explanation of an error constitutes a universal statement that is refutable at the point of error. If the new universal statement is attached to the previous model or takes the place of another universal statement in the previous model, an adjusted model is formed. The adjusted model will be an improvement if it is corroborated at the previous point of error. If the improved model has more universal statements than its predecessor, that is, if variables have been endogenated, then so much the better because the improved model will have even more refutable points for further testing and improvement.

Causal explanations with multiple universal statements involve unobservable relations and hence often require very imaginative testing. Newton's explanation of projectile motion, for example, involved two universal statements. Taken individually, each statement contradicts observed behavior: the gravitational law alone implies that the projectile should fall to the ground, while the inertial principle alone implies that the projectile should move in a straight line. Testing just projectile motion then cannot be adequate for corroborating either of the two universal statements. Instead, there must be a search for situations in which just one of the statements is required (such as explaining free-falling bodies), or in which different combinations of the two statements are required (such as in explaining pendulum motion or planetary motion), *or* in which "thought experiments" can carry out the consequences of the universal statements. Thought experiments can often turn up inconsistencies by tracing a model through extreme conditions that would never actually occur.

Finally, the purpose of testing is *to expose error and to identify its source*. Any test that can do one or the other should be pursued, no matter how absurd it may seem to someone not familiar with the application of refutationist method.

Note

1. Students of Newton's life and work do not believe that he used induction to generate his ideas. His law of gravitation far transcended observable instances from which the law was supposedly induced. Newton talked of gravitational forces between entities that man had never seen and might never see. Furthermore, use of Bacon's legitimation criterion—ideas are scientific if they are reducible to facts—would have rendered Newton's law of gravitation nonscientific. Gravitation cannot be touched or seen, and it has the strange quality of acting at a distance.

References

Agassi, Joseph. 1968. *The Continuing Revolution: A History of Physics from the Greeks to Einstein.* New York: McGraw-Hill.

Bacon, Francis. 1937. *Essays, Advancement of Learning, New Atlantis, and Other Pieces.* Edited by Richard F. Jones. Garden City, N.Y.: Doubleday Co., 1937.

Bacon, Frances. 1960. *The New Organum and Related Writings.* Edited by Fulton H. Anderson. Indianapolis: Bobbs-Merrill Co., 1960.

Bayes, Thomas. 1963. "Essay Towards Solving A Problem in the Doctrine of Chances." In *Facsimiles of Two Papers by Bayes.* New York: Hafner Publishing Co.

Gellner, Ernest. 1974. *Legitimation of Belief.* New York: Cambridge University Press.

Hume, David. 1958. *A Treatise of Human Nature.* Edited by L. A. Selby-Biggs. Oxford, England: Clarendon Press.

Jarvie, Ian. 1964. *The Revolution in Anthropology.* New York: Humanities Press.

Kuhn, Thomas. 1970. *The Structure of Scientific Revolutions.* 2d ed. Chicago: The University of Chicago Press.

Kuhn, Thomas. 1970. "Logic of Discovery or Psychology of Research?" and "Reflections on My Critics." In *Criticism and the Growth of Knowledge*. Edited by Imre Lakatos and Alan Musgrave. Cambridge, England: Cambridge University Press.

Lakatos, Imre, and Alan Musgrave, eds. 1963. *Criticism and the Growth of Knowledge*. Cambridge, England: Cambridge University Press, 1970.

Popper, Karl Raimund. 1963. *Conjectures and Refutations: The Growth of Scientific Knowledge*. New York: Harper and Row Co.

Popper, Karl Raimund. 1970. "Normal Science and its Dangers." In *Criticism and the Growth of Knowledge*. Edited by Imre Lakatos and Alan Musgrave. Cambridge, England: Cambridge University Press.

Popper, Karl Raimund. 1972. *Objective Knowledge: An Evolutionary Approach*. Oxford: Clarendon Press.

Popper, Karl Raimund. 1976. *The Logic of Scientific Discovery*. Rev. ed. New York: Harper-Row.

Toulmin, Stephen. 1970. "Does the Distinction Between Normal and Revolutionary Science Exist." In *Criticism and the Growth of Knowledge*. Edited by Imre Lakatos and Alan Musgrave. Cambridge, England: Cambridge University Press.

Whewell, William. 1847. *The Philosophy of the Inductive Sciences*. 2nd ed. New York: Johnson Reprint Corp., 1967.

Whewell, William. 1858. *History of the Inductive Sciences*. New York: P. Appleton and Co.

The Unavoidable A Priori[1]

2

Donella H. Meadows

2.1 Introduction

Questions are necessarily prior to answers, and no answers are conceivable that are not answers to questions. A "purely factual" study—observation of a segment of social reality with no preconceptions—is not possible; it could only lead to a chaotic accumulation of meaningless impressions. Even the savage has his selective preconceptions by which he can organize, interpret, and give meaning to his experiences.—Myrdal (1968, p. 24)

Although the field of computer modeling has existed for only a few decades, a number of different methodological schools based on distinct procedures and techniques have already appeared. They include linear programming, input-output analysis, econometrics, stochastic simulation, and system dynamics. All these modeling schools share a number of common concepts about the properties of real-world systems, the process of modeling, the use of the computer, and the role of models in decision making.

In addition to the shared concepts general to all mathematical modeling, each methodological school also employs its own special set of theories, mathematical techniques, languages, and accepted procedures for constructing and testing models. Each modeling discipline depends on unique underlying and often unstated assumptions; that is, each modeling method is itself based on a model of how modeling should be done.

These deep, implicit operating assumptions at the foundation of each modeling method are sufficiently important that they should be reexamined more often than they actually are. Practitioners of each method learn its operating assumptions once, and thereafter may reflect on them only occasionally. Typically, the assumptions of each modeling school are part of the subconscious rather than the conscious reasoning that goes into the making of models. Just as physicists rarely rethink the laws of algebra or the second law of thermodynamics as they work, practicing econometricians seldom stop to question their use of statistics to measure model validity, and system dynamicists regularly use the principle of feedback control without redefining it each time.

Different modeling world views, or in Thomas Kuhn's terminology, paradigms (Kuhn, 1970), cause their practitioners to define different problems, follow different procedures, and use different criteria to evaluate their results. In a very real sense the paradigm biases the way the modeler sees the world, and thus influences the content and shape of his models. As Abraham Maslow says, "If the only tool you have is a hammer, you tend to treat everything as if it were a nail." (Maslow, 1966) The selective blindness induced by any operational paradigm has both unfortunate and fortunate results. Unfortunately, it often leads to sterile arguments across paradigms, each school criticizing the problems, assumptions, and standards of the other from the biased perspective of its own problems, assumptions, and standards. On the other hand, paradigm-directed research, which Kuhn calls "normal science," seems to be both psychologically necessary—our minds cannot operate without some selection and categorization of incoming sensations—and also exceptionally fruitful. "Within those areas to which the paradigm directs the attention of the group, normal science leads to a detail of information and to a precision of the observation-theory match that could be achieved in no other way" (Kuhn, 1970, pp. 64–65).

Because of the inescapable effect of methodological paradigms on modelers' thoughts and perceptions, any comparison or evaluation of models must begin with an understanding of the underlying paradigms within which models are made. Furthermore, in order for social system modeling to produce a cumulative understanding of social processes and contribute significantly to social policy making, the problems of selective blindness and of cross-paradigm communication must be dealt with. Computer modeling would be more effective, both as a science and as a useful art, if each modeler could recognize the assumptions behind his own modeling school and could understand and respect the assumptions behind other schools.

To become actively aware of the deep and implicit operating assumptions that guide one's daily professional activities is a surprisingly difficult task. It is even more difficult to discover someone else's operating assumptions, since they are never directly observable and are often antithetical to one's own habitual way of viewing the world. The rarity of this kind of paradigmatic overview is evident from the misunderstandings and even occasional hostility that various kinds of modelers exhibit toward each other, and the distrust that other professionals sometimes exhibit toward modelers in general.

This chapter attempts to expose the bedrock assumptions that underlie the entire field of social system modeling and the more specific assumptions that define two modeling schools—system dynamics and econometrics. My viewpoint here must necessarily be influenced by my own background as a onetime physical scientist turned system dynamicist, one who is relatively new to that field and who has theoretical knowledge of, but little practical experience with, econometrics.

The biases associated with this viewpoint will be readily apparent to everyone but me. Anyone who undertakes the task of defining paradigms will bring some set of biases to it. Mine will serve as well as anyone's to begin the discussion and to bring forth the clarifications and rebuttals necessary to produce a balanced view.

2.2 The Preconceptions of Modeling

Although modelers may disagree vehemently about their specific methods or models, they are unified by some very basic assumptions that define the whole modeling approach to problem solving. First of all, social system modelers generally come from or were educated in a western culture, where attempts at a rational, logical, scientific mode of thought predominate. Whatever happens is not believed to be random; it is assumed to have a cause that can be understood and probably altered. Careful measurement, clever experimentation, and logical deduction should reveal that cause. Intuition and emotions are to be suppressed.

Furthermore, modelers share a basically managerial world view. Problems should be actively solved, not passively endured. One does not ride along with the process of social evolution, one strives to direct that evolution. This managerial world view is generally acceptable to engineers, businessmen, some scientists, and some politicians but not to most artists, theologians, or other humanists, or to people educated in traditional eastern cultures:

> By one Chinese view of time, the future is behind you, where you cannot see it. The past is before you, below you, where you can examine it. Man's position in time is that of a person sitting beside a river, facing always downstream as he watches the water flow past. . . .

> In America and other western countries, the commonest view of abstract time seems to be the opposite of the old Chinese one. In this, man faces in the other direction, with his back to the past, which is sinking behind him, and his face is turned upward to the future, which is floating down upon him. Nor can this man be static: by our ambitious western convention, he is supposed to be rising into the future under his own power, perhaps by his own direction. He is more like a man in a plane than a sitter by a river. (Peck, 1967, pp. 7–8)

Although computer modelers use historic observations to help form their hypotheses, their faces are primarily turned upstream, toward the future. They believe that the future can and should be shaped by decisions and actions based on scientific understanding.

The assumptions that distinguish computer modelers from other managerially oriented scientists center around the tools modelers choose to help them analyze problems: mathematics and the computer. A computer modeler assumes that the computer augments the human brain the way a steam engine augments human muscle, and thus it is the obvious tool for dealing with matters that are too complex for the unaided single mind. Furthermore, he assumes that human actions and purposes can be categorized, quantified, and represented by mathematical equations. This postulate does not necessarily imply, as many nonmodelers believe it does, a belief that human beings or the systems they create are totally predictable. It does require a belief that they are predictable in the aggregate and on the average, however. As E. F. Schumacher says,

> In principle, everything which is immune to the intrusion of human freedom, like the movements of the stars, is predictable, and everything subject to this intrusion is unpredictable. Does that mean that all human actions are unpredictable? No, because most people, most of the time, make no use of their freedom and act purely mechanically. Experience shows that when we are dealing with large numbers of people, many aspects of their behavior are indeed predictable; for out of a large number, at any one time, only a tiny minority are using their power of freedom, and they often do not significantly affect the total outcome. (Schumacher, 1973, p. 217)

Modelers believe not only that aggregate human actions can be quantified into computer equations and that computer equations can be grouped into representations or models of social systems but also that these models are at least potentially better representations than any others that might be used as a basis for social decisions. Jay W. Forrester postulates that individual and social decisions must be made on the basis of some model, most usually the "mental model," which is the set of unexpressed assumptions and generalizations about the world that exists in each person's mind. There is no complete knowledge upon which to base one's actions, there are only more or less simplified models that are derived from education, culture, and personal experience. Given that decisions must be based on some sort of incomplete and uncertain model, a computer model may be preferable to a mental model because

1. it is precise and rigorous instead of ambiguous and unquantified,
2. it is explicit and can be examined by critics for inconsistency or error,
3. it can contain much more information than any single mental model,
4. it can proceed from assumptions to conclusions in a logical, error-free manner,
5. it can easily be altered to represent different assumptions or alternate policies.

Very few computer modelers can claim that their models actually do exhibit all these advantages. Models can easily become so complex that they are impenetrable, unexaminable, and virtually unalterable. They can also be *less* complete than mental models, if requirements of mathematics or data prohibit the inclusion of certain kinds of information. However, these five advantages are considered at least potential characteristics of computer models, and to the extent that they are realized, computer modelers believe they can provide valuable and otherwise unattainable knowledge for social decision making.

2.3 Categories of Modeling Methods

Upon the foundation of these basic assumptions about rationality, the scientific method, the computer, and the advantages of mathematical models, a number of different modeling schools have been erected. Each was originally developed in response to a specific social need, each has developed its own methods and languages, and each shapes the procedures and perceptions of its adherents in a distinct way.

The different kinds of modeling can be classified along a number of dimensions, some of which are partially overlapping and some of which are totally incommensurate. Models may be distinguished by their *information bases* (for example, social statistics or ecological observations), by the *mathematical procedures* they employ (differential equations, simultaneous equations, optimization, and so forth), or by the nature of the model *relationships* (such as stochastic, deterministic, or nonlinear).

I could organize the following discussion around any one of these properties, but instead I shall choose as the primary point of distinction another property that is not very often mentioned in model classifications. That is the *use* to which the model is to be put. I am concerned here only with models that contribute to the understanding and management of social systems. Therefore I shall classify models according to the stage of social decision making at which they are most applicable.

At the very first identification of a problem there may be a need for *general understanding*. The system producing the problem may never have been critically studied, or past studies may have been incomplete or faulty. Important data may be missing, interconnections that had been considered absent or unimportant may suddenly appear significant, or old theories may be called into question by new and unexpected behavior. The problem must be understood in its long-term historical perspective and in a wide enough boundary to include all of its causes and consequences. Current models, mental or otherwise may need revision, updating, or complete overhaul before the problem can be tackled. Examples of social problems in this stage might include promotion of economic development, the transition from petroleum to new energy sources, or control of global pollution. Models that can contribute to an improved general understanding must be easily understood. They should make clear exactly how their assumptions lead to their conclusions, and they should provide new insights about the working of the real-world system. Quantitative precision is unnecessary and probably unattainable at this point; it is difficult enough to decide what system elements are even qualitatively important and how they are related. General-understanding modeling projects tend to be more *process oriented* than product oriented; that is, the very process of making the model, asking questions systematically, and defining new conceptions may itself improve understanding so much that, by the time the computer model is finished, it is no longer needed. Its concepts and conclusions have been integrated into the mental models of both modelers and clients.

If general understanding allows some agreement about the cause of the problem or the nature of the system generating it, then the second phase, which I will call *policy formulation*, begins. Theories about the cause of the problem will lead directly to suggestions about the general directions in which a cure might be found. Broad policy choices must be evaluated and compared to identify possible trade-offs or synergies. The policy questions to be answered by a model are still imprecise and generic at this stage, but the examination can be limited to those points in the system that have been identified as potential policy foci. Should family planning or health care be given more emphasis in aid programs? Should basic research funds be allocated more to solar energy or fusion energy? What would be the general consequences of a ban on DDT? A model that can help in policy formulation should be able to reproduce the real system's behavior under a variety of conditions, it should be easily altered to test a wide variety of possible policies, and it should clarify why different policies lead to different results. Quantitative precision is more important here than it was at the level of general understanding, but the emphasis is still primarily qualitative and process oriented.

When a basic policy direction has been chosen, a whole host of new questions arise concerning the *detailed implementation* required to carry out that policy. A policy to promote family planning engenders numerous further decisions about budgets, personnel training, geographic distribution, and educational techniques. A policy to stabilize grain prices by creating a buffer stock will require the creation of new organizations to establish and maintain the stock, a precise set of rules for buying and selling, and a plan linking markets, warehouses, transportation systems, and final consumers so that the greatest stabilization can be realized at the least cost.

These detailed implementation decisions may be complex and require the organization and processing of many pieces of information. Mathematical models are ideally suited to such tasks. Implementation stage models typically are detailed and highly accurate, but each needs to represent only one basic policy direction, so its boundary can be narrow. Detailed implementation-modeling methods are usually *product oriented*; they are often difficult, expensive, and tedious to construct, but once completed, they can be used over again to transform new input data into specific predictions or operating instructions. Product-oriented modelers rarely involve the client in the modeling process or try to make clear all the model's assumptions. Probably most computer models now being made are directed to this stage of detailed decision making.

Different people sit at the various stages of the policy process, asking different sorts of questions that require different kinds of models. Each of the modeling paradigms can be regarded as a useful tool in a tool box. Knowledge of all their properties is essential in deciding which is the best tool for a given specific purpose. It is possible to use each of these methods for several different purposes, and, by stretching things a bit, for all purposes. A saw could be used to pound in a nail, if necessary. But a hammer does the job better and faster. An essential aspect of wisdom in making, sponsoring, promoting, or criticizing models is knowing when each kind of model is most useful and when it is being pushed beyond its range of applicability; such insight requires a comprehensive understanding of the characteristics, strengths, and weaknesses of each modeling school.

2.4 Two Modeling Schools

I will describe two modeling methods here: system dynamics and econometrics. After a brief summary of the historical development of each field, the most important characteristics and assumptions of the method will be discussed. Examples of actual policy applications will be given. Finally, the most common problems and limitations of the method will be described. These problems are not necessarily present in all models; in fact, the best models are often recognized as good because they have managed to avoid them. Nevertheless, every method has its most common pitfalls, into which students often fall and against which advanced modelers must continually guard. Understanding these potential limitations as well as the strengths of the various modeling schools may be one of the most effective steps toward better modeling and better use of models.

General Characteristics of System Dynamics

System dynamics was developed at MIT during the 1950s, primarily by Jay W. Forrester. He brought together ideas from three fields that were then relatively new—control engineering (the concepts of feedback and system self-regulation), cybernetics (the nature of information and its role in control systems), and organizational theory (the structure of human organizations and the forms of human decision making). From these basic ideas, Forrester developed a guiding philosophy and a set of representational techniques for simulating complex, nonlinear, multiloop feedback systems. He originally applied these techniques to problems of industrial firms. The first system dynamics models addressed such general management problems as inventory fluctuations, instability of labor force, and falling market share (see Forrester, 1961).

The methods worked out by Forrester and his group have since been applied to a wide variety of social systems (see, for example, Forrester, 1968, Hamilton et al., 1968, and Forrester, 1971). The field is still dominated by engineers, industrial managers, and physical scientists, all of whom are generally more interested in the application of the method than in its theory. The literature of system dynamics contains many more descriptions of models addressed to policy questions than analytical discussions of modeling techniques.

As its name implies, system dynamics is a method of dealing with questions about the dynamic tendencies of complex systems, that is, the behavioral patterns they generate over time. System dynamicists are generally unconcerned with precise numerical values of system variables in specific years. They are much more interested in general dynamic tendencies; whether the system as a whole is stable or unstable, oscillating, growing, declining, or in equilibrium. To explore the dynamic tendencies of systems, they will include in their models concepts from any discipline or field of thought, with special emphasis, however, on the physical and biological sciences and some tendency to discount (or rediscover and rename) theories from the social sciences.

The primary assumption of the system dynamics paradigm is that the persistent dynamic tendencies of any complex system arise from its *causal structure*—from physical constraints and social goals, rewards and pressures that cause people to behave the way they do and to generate cumulatively the dominant dynamic tendencies of the total system. A system dynamicist is likely to look for explanations of recurring long-term social problems within this internal structure rather than in external disturbances or random events. For example, a system dynamicist is led by his paradigm to explain the U.S. energy problem in terms of the habits, constraints, aspirations, and decisions of the American people, rather than as a result of Arab oil embargoes or bad weather. He is likely to look for a solution to the problem in changing the goals and the information that influence people's decisions, not through small adjustments in taxes, research expenditures, environmental standards, or foreign policy. This basic assumption does not necessarily imply that all problems originate from faulty system structure—just that system dynamicists are more likely to see and become interested in the problems that do.

The central concept that system dynamicists use to understand system structure is the idea of two-way causation or feedback. It is assumed that social or individual decisions are derived from information about the state of the system or environment surrounding the decision-maker. The decisions lead to actions that are intended to change the state of the system. New information about the changed state (or unchanged, if the action has been ineffective) then produces further decisions and changes. Each such closed chain of causal relationships forms a feedback loop. System dynamics models are made up of many such loops linked together. They are basically closed system representations; most of the variables occur in feedback relationships and are thus endogenous (that is, determined within the model system). Relatively few variables are represented exogenously (influence the system but are not influenced by it).

The element in each feedback loop that represents the environment surrounding the decision-maker is referred to as a state variable or *level*. Each level is an accumulation or stock of material or information. Typical levels are population, capital stock, inventories, and perceptions, which are accumulations of information. The element representing the decision, action, or change (often, but not always, induced by human decision-makers) is called a *rate*. A rate is a flow of material or information to or from a level. Examples are birth rate, death rate, investment rate, or rate of sales from inventory.

The concepts of feedback, levels, and rates require a careful distinction between stocks and flows of real physical quantities and information. In the system dynamics paradigm physical flows are constrained to obey physical laws such as conservation of mass and energy. Information flows obey their own particular laws: information need not be conserved, it may be at more than one place at the same time, it cannot be acted upon at the same moment it is being generated, it may be systematically biased, delayed, amplified, or attenuated.

Two kinds of feedback loops are distinguished. A positive, or self-reinforcing, loop tends to amplify any disturbance and to produce exponential growth. A negative, or goal-seeking, loop tends to counteract any disturbance and to move the system toward an equilibrium point or goal. Certain combinations of these two kinds of loops recur frequently and allow system dynamicists to formulate a number of useful generalizations or theorems relating the structure of a system (the pattern of interlocking feedback loops) to the system's dynamic behavioral tendencies. For example, when the system exhibits exponential growth, the system dynamicist automatically looks for a dominant positive feedback loop. A tendency for a system to return to its original state after a disturbance indicates the presence of at least one strong negative feedback loop. Oscillatory behavior often indicates the presence of a negative feedback loop with a time delay in it. Sigmoid or S-shaped growth results from linked positive and negative loops that respond to each other nonlinearly and with no significant time delays.

These and other structure-behavior theorems are the main intuitive guides that help a system dynamicist interpret the observed dynamic behavior of a real-world system and detect structural insufficiencies in a model. They permit identification of isomorphisms in very different systems that can be expected to have similar behavior patterns. For example, to a system dynamicist, a population with birth and death rates is structurally and behaviorally the same as an industrial capital system with investment and depreciation rates. They each consist of linked positive and negative loops with time delays, and from their structure they can be expected to grow exponentially, decline exponentially, or oscillate, but not to exhibit sigmoid growth (because of the time delays).

As these examples illustrate, time delays can be crucial determinants of the dynamic behavior of a system. System dynamics theory emphasizes the characteristics and consequences of different types of delays, both in information and in physical flows. System dynamicists expect and look for lagged relationships in real systems.

Nonlinearities are also believed to be important in explaining system behavior. A nonlinear relationship causes the feedback loop of which it is a part to vary in strength, depending on the state of the system. Linked nonlinear feedback loops thus form patterns of shifting loop dominance —under some conditions one part of the system is very active, and under other conditions another set of relationships takes control and shifts the entire system behavior. A model composed of several feedback loops linked nonlinearly can produce a wide variety of complex behavior patterns.

Nonlinear, lagged feedback relationships are notoriously difficult to handle mathematically. Forrester and his associates developed a computer simulation language called DYNAMO that allows nonlinearities and time delays to be represented with great ease, even by persons with limited mathematical training. DYNAMO is a very specialized language developed to express the basic postulates of the system dynamics paradigm and to be easily understandable to laymen. It is widely used by system dynamicists because of its clarity and convenience, and therefore it is often thought to be an identifying characteristic of a system dynamics model. But any system dynamics model can be written in a general purpose language such as FORTRAN, and, conversely, DYNAMO can be used to program linear open-system models that are not philosophically system dynamics models at all. In other words, DYNAMO is a tool used by many system dynamicists, but it is not exclusively a system dynamics tool.

System dynamics models are usually used at the general-understanding stage of decision making. Therefore they tend to be fairly small and transparent. Most fall within the range of 20 to 200 endogenous variables. The individual model relationships are usually derived directly from mental models and thus are intuitive and easily understandable. The paradigm requires that every element and relationship in a model have a readily identifiable real-world counterpart; nothing should be added for mathematical convenience or historical fit. Great emphasis is placed on careful model documentation and on involving the client as much as possible in the modeling process.

Some questions that have been addressed with system dynamics models include:

- What has caused American cities to experience a 100 to 200 year life cycle of growth, followed by stagnation and decay? (Forrester, 1969)
- How do primitive slash-and-burn agricultural societies control their populations and their land use practices to ensure a stable pattern of life in an ecologically fragile environment? (Schantzis and Behrens, 1973)
- Why do agricultural and mineral commodities exhibit oscillating price and production trends, and why does each commodity exhibit a characteristic period of oscillation? (Meadows, 1970)
- What has caused the decrease in the number of economically viable dairy farms in Vermont, and what policies might halt that decrease? (Budzik, 1975)
- What policies will help the U.S. energy system make a smooth transition from a petroleum base to other energy sources? (Naill et al., 1975)

The first three of these studies fall in the category of general understanding; the fourth and fifth include both general-understanding and policy formation. All of the studies have a time horizon of thirty years or more.

Problems and Limitations of System Dynamics

System dynamics modelers, particularly when using the DYNAMO compiler, must supply knowledge and judgment about interconnections in the real-world system, but not extraordinary mathematical or programing skill. The DYNAMO software package has many obvious advantages, but it also has several disadvantages. First, it makes the mechanics of modeling so easy that beginners who know the language but not the underlying philosophy of the method are likely to become overconfident and to oversell their skills and their models. Second, because additions, alterations, and policies are readily added and analyzed within minutes, beginners and advanced modelers alike are tempted to play with endless model variations, rather than to think through carefully the experiments they have tried and the lessons they have learned. Finally, the simplicity of adding new elements and relationships to a model encourages the natural tendency of all modelers to create overcomplex, incomprehensible structures.

The ease with which models can be overelaborated is common to many modeling schools but is a special problem in system dynamics. Both the philosophy and the general-understanding purpose of the system dynamics method require simplicity and transparency. System dynamicists recognize the problem of overcomplex models and greatly emphasize, both in training and in publication, the necessity and difficulty of creating simple models. System dynamicists tend instinctively to criticize complex models and to admire simple ones. In fact, the pains that are taken to instill and reiterate the goal of model simplicity reflect the very real difficulties in achieving it.

The emphasis on simplicity in system dynamics is consistent with the purposes for which this technique is usually intended, but it has also limited its range of application primarily to questions that involve aggregate quantities rather than questions of distribution. Distribution of income, resources, opportunity, pollution, or any other quantity is represented in almost any modeling method by the "brute force" method of disaggregation. Each class, person, or geographical area concerned is represented explicitly, and the flows of goods or bads among them is accounted for. Disaggregation into even a few interacting classes or levels can complicate a model tremendously. A modeler striving for clarity will try to avoid disaggregation as much as possible, and thus may be likely to discount or simply not perceive questions of distribution. This does not mean that system dynamicists are unable to deal with distribution questions, but their paradigm gives them a certain reluctance to disaggregate.

Three problems that recur in all modeling techniques but that are relatively less bothersome in system dynamics than in other modeling schools are estimation of parameters, sensitivity testing, and assessment of model validity.

Statistical estimation procedures are seldom used in system dynamics for four reasons. First, most system dynamics models are not directed to problems of detailed implementation or precise prediction, but to problems of general understanding that do not require highly accurate numbers. Second, because of the long-term nature of most system dynamics problem statements, parameters are likely to exceed historic ranges, so estimation based on historic data alone would be insufficient. Third, the nonlinear feedback structure of the models makes standard statistical techniques either inapplicable or extremely difficult. And fourth, the nonlinear feedback structure of system dynamics models renders them less sensitive to precise refinements of parameter values.

The general insensitivity of system dynamics models is partly a result of their feedback structure, but it is also partly due to the way sensitivity is defined in the system dynamics paradigm. Model output is read not for quantitative predictions of particular variables in particular years but for qualitative behavioral characteristics. A model is said to be sensitive to a given parameter only if a change in the numerical value of the parameter changes the entire behavior of the model (from growth to decline, for example, or from damped oscillation to exploding oscillation). Sensitivity of this kind is extremely rare, both in system dynamics models and in social systems, but it does occasionally occur. In fact, detection of a particularly sensitive parameter is an important result of the modeling process, because it earmarks that parameter as one that must be estimated carefully or one that might be an effective site for policy input.

No rigorous theory or procedure exists in system dynamics for performing sensitivity analysis, and this is a weakness of the field. On the other hand, the informal structure-behavior theorems that characterize the paradigm sometimes permit an experienced dynamicist to locate sensitive parameters by inspection of the model structure and thus to elimate the necessity of testing every possible parameter in the system. This intuitive approach to sensitivity testing is very effective in small models but unusable in large ones.

The system dynamics paradigm handles the problem of model validity qualitatively and informally. There is no precise, quantitative index to summarize the validity of a system dynamics model. In fact, system dynamicists do not usually use the term validity. Reference is made to model *utility*. Is the model sufficiently representative of the real system to answer the question it was designed to answer? A system dynamicist begins to have confidence in his model when it meets the following conditions:

1. Every element and relationship in the model has identifiable real-world meaning and is consistent with whatever measurements or observations are available.
2. When the model is used to simulate historical periods, every variable exhibits the qualitative, and roughly quantitative, behavior that was observed in the real system.
3. When the model is simulated under extreme conditions, the model system's operation is reasonable (physical quantities do not become negative or exceed feasible bounds; impossible behavior modes do not appear).

These standards are imprecise and do not lend themselves to quick evaluation. They are also quite difficult to achieve in practice. The issue of model validity is an unresolved one in every modeling field. System dynamics approaches it by admitting the indeterminancy of the very concept of validity and by establishing performance standards that are qualitative but demanding.

The most difficult problems in system dynamics appear in the process of modeler-client interaction. The system dynamics paradigm leads the analyst naturally to a long-time horizon and wide-boundary approach to any problem. This viewpoint is not usually consistent with the very real short-term pressures and constraints felt by most decision makers. The result may be an impasse; the client cannot take the broad perspective of the modeler, and the modeler is convinced that no other perspective will lead to a problem solution.

Since system dynamicists assume that most problems, like most model elements, are endogenous to the system, they will look for and often find internal decisions to be a major cause of observed problems. The recommended solution often requires structural change. This change may be as simple as bringing new information to bear on a decision, but it may also involve revision of goals, reward structures, or areas of authority. These recommendations are often politically unacceptable. This problem is intrinsic to the basic paradigm of system dynamics and the nature of public decision making and will probably always be a factor hindering the practical use of system dynamics in the policy world.

General Characteristics of Econometrics

Econometrics is a widely practiced and more varied field than system dynamics, and no general description can cover the diversity of individual practitioners. In the following description I attempt to capture the common characteristics of the majority of econometric models, the field as it is in practice, not in theory, and, of course, the field as it is seen by an outside observer. In discussing the weaknesses of econometrics, I have made use of quotations from practicing econometricians, since their criticisms must be more informed and more credible than mine.

Econometrics is defined as the use of statistical methods to verify and quantify economic theory. A set of theoretical relationships that has been verified and quantified for a particular economic system constitutes an econometric model of that system. The model can be used for structural analysis, for forecasting, or for testing the effects of policy alternatives.

The field of econometrics combines tools and concepts from the two older fields of statistics and economics. Therefore it shares aspects of both those paradigms, as well as adding its own special perspectives to the world-view of its practitioners. Statistical economics developed in the 1930s as a result of a rising interest in the quantitative behavior of national economic variables, especially aggregate consumption, which was postulated to be a major cause of the problems of the great depression. The journal *Econometrica* was begun in 1933. By the late 1930s Jan Tinbergen had constructed the first dynamic models of the Dutch, U.S. and British economies (Tinbergen, 1937). Much theoretical and practical work had already been done by the early 1950s, when the development of the computer permitted a great expansion in the scope and complexity of econometric models.

The dominating characteristic of the econometric paradigm is its reliance on statistical verification of model structure and model parameters. Econometricians are forced by their paradigm to tie their models firmly to statistical observations of real-world systems. The formulation of an econometric model may be divided theoretically into two sequential phases: specification of structure from economic theory, and estimation of parameters by statistical analysis. The second phase is the center of concern, however, occupying most of the modeler's time and attention and most of the pages in econometric textbooks and journals. To some extent the mathematical and data requirements of the estimation phase affect the specification phase as well, as will be described later.

The information base from which an econometrician can draw his model structure is the same one underlying system dynamics or any other modeling technique—abstractions, intuitions, personal experiences, statistical data, established wisdom, experimentation, and guesswork. In practice, most econometricians are attracted to questions about the precise, short-term values of economic variables. They find most of the concepts they need in traditional economic theory. They tend to make only limited use of theories from other disciplines, and when they do, their bias tends to be as much toward the social sciences as the system dynamicists' bias is toward the physical sciences. No special distinction is made between the physical and information flows in econometric models. For example, many of the common variables in econometric models are expressed in units of unconserved monetary stocks and flows, even when they stand for conserved physical stocks and flows (examples are production, consumption, capital, investment, depreciation, imports, and exports).

The underlying economic theory from which econometrics is drawn is much richer in static concepts than dynamic ones, perhaps because much of the theory was developed before computer simulation allowed dynamic analysis of complex, nonlinear systems. Much attention is paid in economics to the optimum or equilibrium points in a system, comparatively little to the path of approach to equilibrium or the time required to attain it. Although many econometric models are dynamic, they maintain their parent field's emphasis on optima and equilibria rather than on dynamic characteristics.

Economic theory also leads econometric modelers to create structures that are partially open—driven by many exogenous variables that must be forecast independently from the model— rather than entirely closed into feedback loops that cause the system to move itself through time. Economics evolved as an open-system body of theory for several reasons. Economic systems are strongly driven by forces outside the disciplinary boundary; resources come from the domain of geology, weather fluctations from meteorology, consumer motivations from psychology and sociology, and labor availability from demography. Furthermore, the relatively short-term focus of many economic problem statements means that analysts often need not take into account feedback processes with long-time delays.

When two-way causation does appear in econometric models, it is typically represented by means of simultaneous equations. The simultaneous-equation formulation is equivalent to assuming that system equilibrium will occur within one calculation interval.

Although most econometric models contain simultaneous-equation formulations and are driven dynamically by exogenously forecasted variables, many models also contain some feedback through lagged endogenous variables. These formulations are not essentially different from those in system dynamics models. The distinction between the two approaches is one of relative emphasis, not absolute contrast. Econometric models contain some feedback relationships, some of which are lagged; system dynamics models are composed almost entirely of feedback relationships, all of which are lagged.

Even within the disciplinary boundary of economics, the variables that can be included in econometric models are restricted to a subset of all conceivable elements, because of the necessity for statistical validation. Each element in an econometric model must be observable because sufficient historic observations of it must exist to permit precise estimation of its quantitative relationship to other variables. That requirement tends to eliminate the inclusion of most of what system dynamicists call the information components of any system, especially the motivations behind human decisions. These motivational components are not absent from economic theory, which contains many inherently unobservable concepts such as marginal utility, indifference curves, and the profit motive. But none of these ideas are easily made tangible, and therefore they do not appear as explanatory variables in econometric models.

The requirement of observability is not as confining to econometricians as a system dynamicist might think. In the long run it creates the pressures that are already improving and expanding data collection efforts around the world. Useful, but unmeasurable, concepts eventually can become sufficiently well defined to be measured and included in data bases. For example, no GNP statistics were available until economists devised the concept, found it useful, and figured out how to measure it.

Econometricians can often represent an unobservable concept by means of a closely correlated tangible substitute or proxy. National literacy data may suffice as a stand-in for degree of modernization, rainfall may be a proxy for all the effects of weather on crop production, or advertising expenditures may be used to represent some of the motivational assumptions underlying a consumer demand equation. In other words, an econometrician can transform a direct hypothesis about the causes of behavior such as "in early stages of modernization people's material aspirations rise, therefore they consume less and save more for investment to increase their future consumption," into an indirect hypothesis about correlation of observables, such as "at low income levels literacy is inversely correlated with consumption." The use of correlated rather than direct causally related variables allows econometricians to proceed in spite of the requirement of empirical validation, but it also reinforces that requirement because a double set of assumptions has been made. The original relationship was hypothesized between modernization and consumption, but now a new hypothesis has been added about the correlation between modernization and literacy. Both relationships are tenuous and always subject to change, and therefore they must be rechecked continuously against real-world data.

The principal technique used to obtain parameters for econometric models is least squares estimation, a method that generates the set of numbers that best fits a postulated general relationship to historic observations and that also provides a quantitative measure of how good that fit is. The theoretical and mathematical requirements of this method impose several conditions that cause econometric models to depart from economic theory. For example, it requires that the equations be convertible to a form in which all parameters to be estimated enter linearly. As a consequence, most relationships in an econometric model are linear or log-linear. The assumed relationship between literacy and consumption is most likely to be expressed as

$$\text{consumption} = \beta_0 + \beta_1 \text{ (literacy)} + E,$$

or perhaps as

$$\log \text{ (consumption)} = \beta_0 + \beta_1 \text{ (literacy)} + E,$$

where β_0 and β_1 are constants called *structural coefficients*, to be determined by fitting historical data for consumption and literacy. The "error term" E, measures the observed variation in consumption that cannot be accounted for by variations in literacy.

Another requirement of least squares estimation is that the variation in each explanatory variable must not be linearly dependent on the variation in any other variable and must be strictly independent from the error term. Thus if consumption were postulated to be a function of both literacy and income,

$$\text{consumption} = \beta_0 + \beta_1 \text{ (literacy)} + \beta_2 \text{ (income)} + E,$$

the statistical procedures for estimating β_0, β_1, and β_2 will be accurate only if there is no high degree of correlation between income and literacy or between either of those any of the omitted factors that might influence the error term.

The effect of this requirement on model specification is a subtle psychological one; in order to avoid the numerical biases that result from covariance of variables with each other or with the error term, econometricians tend to include relatively few explanatory variables in their equations:

> By the very existence of a large number of intercorrelations among all economic variables we can estimate but a few partial coefficients with tolerable precision. This accounts for the contrast between economic theory and empirical research. The theory is comprehensive: if we list the determinants of, say consumption or investment that have been discussed by economists, we may easily find some ten or twenty distinct effects. But in econometric research we rarely try to estimate more than four or five coefficients. (Cramer, 1971)

The structural coefficients in an equation like the one relating literacy and consumption are estimated for the system of interest by finding observed values for all variables over some historical period or over some cross-section of subsystems (families, nations, or firms, for example). Ideally, the observed values are used to estimate the structural coefficients of the model, and then the model with its estimated parameter values is used to generate or simulate the values of system variables for another time period or over another cross-sectional sample. The entire procedure depends upon the assumption that the underlying causal mechanisms do not change in form, strength, or stochastic properties from the estimation period to the forecasting period or from one cross-sectional sample to another. Various statistical indices, such as the square of the multiple correlation coefficient R^2, are used to summarize the extent to which the model-generated values can be expected to duplicate the variance in the observed values.

Econometric models tend to deal with highly aggregated quantities, even more aggregated than those in system dynamics models. Typically, few variables are included, and the models are relatively small, consisting of 1 to 200 equations.

In summary, econometricians tend to represent systems as highly linear, partly open, at or near equilibrium, and centered around variables that fall within the disciplinary boundary of economics. The real-world systems that are most congruent with this image encompass flows of economic goods and services, money and prices, over a fairly short time horizon. In these systems many important influences are indeed exogenous, and many relationships are constrained within ranges that are very nearly linear. Also, over the short term the numerical coefficients derived from historical observations are still likely to be valid. If appropriate data are available, econometric methods can provide very precise information about such systems. Thus econometric models are mostly used for short-term predictions of aggregate economic variables. They are least applicable to policy questions that may range across disciplines, over long time horizons, or into circumstances that have not been historically observed.

Examples of questions addressed by recent policy-oriented econometric studies include,

• Will a change in the oil import quotas of the U.S. aggravate the shortage of domestic natural gas from now to 1985, and if so, what wellhead natural gas price would alleviate the shortage? (Spann and Erickson, 1973)
• What will be the effect on U.S. economic growth from now to 2000 if there is more or less government spending, faster or slower population growth, sustained or decreased technical progress? (Hudson and Jorgenson, 1974)
• What will be the quarterly consumer price index for food over the next four quarters? (Barr and Gale, 1973)
• How many acres will be planted in wheat in the U.S. next year if government acreage restrictions and loan programs are altered in various ways? (Hoffman, 1973)
• How will fiscal and monetary control decisions of the Federal Reserve Board affect the U.S. macroeconomy over the next few years? (Modigliani et al., 1973)

Most but not all of these policy questions fall within the short-term, narrowly bounded range of the implementation stage of decision making.

Problems and Limitations of Econometrics

The greatest strength of the econometric paradigm is its insistence on continuous, rigorous checking of theoretical hypotheses against real-world data. This strength leads, however, to two problems already noted: the statistical methods used for estimation impose artifical restrictions on the initial formulation of the model; the data necessary for proper verification are seldom available.

The mathematical requirements of estimation cause econometricians to represent economic systems as linear, mostly simultaneous relationships connecting a few aggregate economic variables by means of historically observed coefficients. My system dynamicist's bias causes me to suspect that real economic systems are nonlinear, multivariable, time-delayed, disaggregate, and ecological-socioeconomic, and they may respond to policy decisions in ways that are not represented in historical data. However, there must certainly be parts of these systems that fit the narrow domain of econometrics quite well. Within these areas econometric techniques can produce accurate, informative, precise, and useful predictions. The major problem in econometric modeling is to recognize the limits of the congruent areas and resist the temptation to push outside them. Thoughtful econometricians know these limits well and seem to conclude that general macroeconomic forecasting purposes are outside the limits:

What then is econometrics . . . best suited for? I myself would place economic problems of the firm in the fore. . . . The problems confronting a firm are in general much less complex than those confronting the economy as a whole: they often are truly of a partial nature. Secondly, the number of observations of the same social system can here frequently be increased. We do not have to face the dilemma of the need for large samples in a world changing rapidly during sampling.

If we wish to use econometrics for macro-economic purposes—to which it first turned, perhaps because disciplines as the outflow of human perversity always first turn to the field of application least suited for them—I would think that it can well be used to test which of a large number of economic hypotheses can best explain an economic reality precisely defined as to time and place. (Streissler, 1970, pp. 73–74)

Nearly every econometrician would list the lack of good empirical data as the most annoying and constricting problem in his field. Econometric researchers pay great attention to data problems and have developed names and categories for the most frequently occurring ones:

Among the more important problems are that there is simply not enough data (*the degrees of freedom problem*); that the data tend to be bunched together (*the multicollinearity problem*); that because changes occur slowly over time, the data from time periods close together tend to be similar (*the serial correlation problem*); that there may be a discontinuous change in the real world so that the data refer to different populations (*the structural change problem*); and that there are many inaccuracies and biases in measuring economic variables (*the errors of measurement problem*). (Intriligator, 1972, p. 157)

Econometric techniques include a number of ingenious methods for recovering from data problems and for extracting maximum possible information from minimal real-world observations. Unfortunately, none of these methods can create more information than is already there, and a process that overcomes one data problem usually makes another one worse:

For example, replacing annual data by quarterly data increases the number of data points but tends to aggravate both the multicollinearity and the serial correlation problems; eliminating data points referring to unusual periods, such as during war years, overcomes the structural change problem but aggravates both the degrees of freedom and the multicollinearity problems; and replacing variables by their first differences overcomes the serial correlation problem but aggravates the errors of measurement problem. (Intriligator, 1972, p.157)

Another criticism econometricians commonly voice about their own field is that econometric modeling is often done badly. In part this may be a by-product of widespread use of econometrics and of very convenient computer software—the same mixed blessing we have encountered in system dynamics. In the case of econometrics, the statistical packages that are now standard equipment at most computing centers can be used rather easily by skilled analysts, and also by those who have never understood or who have entirely forgotten the assumptions underlying the regression techniques. The result can be a blind manipulation of data and an overconfident belief in computed results. Mechanical application of statistical techniques may be substituted for experience with the real-world system, for knowledge of economic theory, and for thoughtful evaluation of conclusions. This is not an insurmountable problem; it can be overcome by better training of modelers, better self-regulation of the econometric profession as a whole, and continuous questioning and review of econometric modeling efforts by modelers, clients, and sponsors.

Because econometric models are partially open systems, they tend to be more sensitive to parameter variation than are system dynamics models. The difference in sensitivity is magnified by the fact that econometricians are usually striving for much more precise statements about the future than are system dynamicists. One would expect, therefore, that sensitivity analysis would be a central concern of econometricians. However, the procedures for carrying out and reporting sensitivity tests, especially for alternate forecasts of the usually numerous exogenous variables, do not seem to be formalized or regularly reported in model documentation. Testing every believable combination of values for exogenous variables would be an impossible task, and the intuitive structure-based hunches that system dynamicists use to detect sensitive points are less applicable to econometrics.

Econometricians determine the validity of their models by the use of statistical tests of model-generated data against real-world data and by the informal comparison of model results with their mental models of "reasonable" values for economic variables. These two validity tests are probably as good as any other when the statistical tests are done honestly and skillfully, and when the modeler has a deep understanding of the workings of real economies. A less honest, skillful, or knowledgeable modeler, however, can produce with these tests evidence of validity for almost any model. In other words, although econometrics techniques include a number of sophisticated statistical validity tests, establishing confidence in a model's output is as difficult and uncertain in this modeling school as it is in all others.

2.5 Paradigm Conflict

The two modeling techniques discussed here are complementary in several ways. For example, system dynamics provides a theory of causal structure, and its relation to dynamic behavior, that is a powerful guide to model specification. Econometrics offers numerous techniques for finding empirical parameters and for formal comparison of model results with real-world observations. One technique is particularly applicable to long-term analysis of possible changes in historic trends. The other is best suited to short-term precise prediction in situations that do not differ from the historic. It would seem that use of the two methods together might produce models that combine realistic structure with accurate parameters, models that are useful at every stage of the decision-making process, particularly for middle-term problems which are not easily analyzed by either method alone.

Unfortunately, this logical combination of two complementary modeling tools has not often been used. On the contrary, very few econometricians have bothered to learn system dynamics techniques, and those system dynamicists who have been schooled in econometrics do not regularly use its tools or concepts. Members of the two schools seem to regard each other as competitors rather than as potential collaborators, and find little to praise in each other's work.

In part this hostility may be due to the personalities of the methodological founders, the natural parochialism of academics, and inevitable jockeying for scarce funding resources. However, a closer examination of the two modeling paradigms reveals a deeper division, one that is not easily bridged. The basic world views upon which the two paradigms are built are quite different, as if they cut through reality with two perpendicular planes that only meet along one narrow line. Either paradigm, seen from the perspective of the other, looks unrealistic and misleading. Methodological conversations between econometricians and system dynamicists tend to degenerate into classic cross-paradigm confusions. Key words such as "validation," sensitivity," and "prediction" are used in different ways based on different implicit assumptions.

Thomas Kuhn is not optimistic about building bridges across paradigm gaps:

The proponents of competing paradigms are always at least slightly at cross-purposes. Neither side will grant all the non-empirical assumptions that the other needs in order to make its case. . . . Though each may hope to convert the other to his way of seeing his science and its problems, neither may hope to prove his case. The competition between paradigms is not the sort of battle that can be resolved by proofs . . . .

The proponents of competing paradigms will often disagree about the list of problems that any candidate for paradigm must resolve. Their standards or their definitions of science are not the same. . . . Communication across the revolutionary divide is inevitably partial. . . . In a sense that I am unable to explicate further, the proponents of competing paradigms practice their trades in different worlds. . . . The two groups of scientists see different things when they look from the same point in the same direction. . . . That is why a law that cannot even be demonstrated to one group of scientists may occasionally seem intuitively obvious to another. Equally, it is why, before they can hope to communicate fully, one group or the other must experience the conversion that we have been calling a paradigm shift. Just because it is a transition between incommensurables, the transition between competing paradigms cannot be made a step at a time, forced by logic and neutral experience. Like the Gestalt switch, it must occur all at once . . . or not at all. (Kuhn, 1970, pp. 148–151)

I believe and hope that the paradigms of system dynamics and econometrics are not as totally incommensurable as Kuhn implies. But there are certainly serious cross-paradigm translation problems that interfere greatly with attempts at synthesis. In this section I will look at both paradigms simultaneously, switching back and forth to see each from the point of view of the other. The resulting image will necessarily be a bit disjointed, since it will not have a constant reference point. It will also magnify the methodological division somewhat, because this description itself is an oversimplified model of reality. And, needless to say, it will not be a totally unbiased description, despite my efforts to make it so. If the following discussion does not induce mutual understanding in the "proponents of the competing paradigms," perhaps it will at least give uninvolved observers of the competition some idea of what each side is assuming, beneath what it is saying.

What is the Problem?

As Kuhn says, the problem begins with the choice of a solvable problem. System dynamicists and econometricians are led by their paradigms to notice different problems and to strive for different kinds of insights into socioeconomic systems.

Econometricians seem to feel that useful information must be detailed and precise—a picture that is not entirely in focus is not worth looking at. They see little substance in the ambiguous, qualitative, long-term output of system dynamics models. To achieve as much precision as possible, econometricians work with statistical methods, which require historic data bases, linear equations, and open structures. They develop little structure-behavior intuition, and they feel that the long term is simply inaccessible to modelers. The lenses they use to look at the world are microscopic, not telescopic, and therefore they conclude that attempts to form a clear image of a happening far away can only be a waste of time.

System dynamicists regard any effort to gain precise predictions of social systems as hopelessly naive. They regard human unpredictability as too dominant a factor in social systems to allow anything more than qualitative behavioral forecasts, even for aggregate systems where much unpredictability can be averaged out. Therefore they find it hard to understand the great effort econometricians go through to obtain better and better estimations or to quote their findings to six or seven significant digits. Especially when many exogenous variables must be predetermined, the whole econometric exercise looks to a system dynamicist like a transformation of one set of uncertain and unscientific guesses into a second set of equally uncertain guesses, presented with deceptive, scientific-looking precision.

System dynamicists should know from their own theorems of system behavior that most aggregate systems possess significant momentum, and that within a short time horizon the relatively simple structural hypotheses of econometrics are usually quite appropriate. But the system dynamics paradigm tends to reject not only the possibility but the utility of working within short time horizons. In the system dynamics world view the short term is already determined and thus unchangeable by policy. Furthermore, in system dynamics models policies designed only for short-term gain often lead to long-term loss.

These different ideas about what kinds of knowledge about the future are useful arise from basically different assumptions about the nature of social systems. The econometric assumption reflects the common view of the policy-making world that the world is essentially dualistic and open. There is a sharp distinction between the economy and the environment (government, weather, Arab nations, consumers, investors, or whatever). The environment delivers specific inputs to which the system gives specific responses. Each system, input, and response may be unique, and thus particulars of different situations are more to be studied than similarities. The best strategy for policy is to foresee the next set of specific inputs and be prepared to give optimal responses to them. This view leads to policy questions about end states, rather than paths to those states, and about particular characteristics of the system under particular conditions:

- If the price of natural gas is deregulated this year, what will its equilibrium market price be? How much windfall profit would accrue to the gas companies?
- How much increase in income taxes would be required to reduce the current rate of inflation by 2 percent? What would that tax do to the unemployment rate?
- Given normal weather conditions, current fertilizer prices, and a subsidy of 5 cents per bushel, how much wheat will be produced in the United States next year? If no export embargoes are imposed, what will domestic wheat price be?

System dynamics, on the other hand, assumes that systems are primarily closed; not only does the environment influence them, but they influence the environment. In fact, the distinction between the system and its environment is rarely clear (except for obvious exogenous factors like incoming solar energy). Attention is focused on the general system reaction to general disturbances and on the dynamic path of a response rather than its end state:

> System dynamics . . . regards external forces as *there,* but beyond control and hence not worthy of primary attention . . . Instead, the focus is upon examining the organization's *internal structure;* the intent being to arrive at an understanding of how this structure. . . . can be made more *resilient* to environmental perturbation. In adopting this approach, system dynamics is embracing the wisdom of the human body. The body, rather than forecasting—and then marshalling its forces in anticipation of—the arrival of each kind of solid and liquid input, remains continually poised in a state of *general readiness* for whatever may befall it. (Richmond, 1976)

This assumption about the nature of systems would lead to a very different set of policy questions:

- How would deregulation of natural gas price affect the general depletion life cycle pattern of U.S. natural gas reserves?
- What are the dominant positive feedback loops causing inflation? How could equally effective negative loops to counterbalance them be built into the economic system without causing unemployment?
- Why has wheat production fluctuated more in the past five years than in the preceding fifteen years? Which sort of policy, direct price supports, increased buffer stocks, or increased exports could induce stabilization of production while not increasing consumer prices?

What Can You Know and How Can You Know It?

After choosing different problems and dismissing the legitimacy and feasibility of each other's problem areas, econometricians and system dynamicists go on to solve their problems with totally different procedures. The differences here have deep roots in conflicting theories of knowledge. Perhaps both sides would agree that the nature of the world and our perceptions of it produce a number of observable happenings that result from an underlayer of unseen causal motivations, events, and connections. The disagreement begins in deciding which part of the double-layered world to represent in a model.

Econometrics is firmly grounded in observable reality. Econometricians may speculate freely about unseen psychological and physical driving machinery, drawing on substantial causal theory from their parent paradigm of economics. But their models must contain explicitly not what they guess but only what they know, and in their paradigm one can know only what one can measure. Therefore their models tend to represent surface phenomena only, with the underlying causal structure implicit. There are no strong preconceived notions about the nature of that structure. It may be an interconnected web of feedback loops; it may be a series of unrelated stochastic forces; or it may be some combination of these. Whatever the underlying structure is, its nature and its relationship to the surface phenomena may change; therefore stochastic error terms must be added to equations, and a continuous stream of new observations must be obtained to verify that the system continues to run as it has in the past. Econometricians therefore feel a pressing need for more data, better measurements, more recent figures, better access to data bases.

System dynamicists, on the other hand, feel that statistical data represent only a small fraction of what one can know. They plunge enthusiastically into the lower layer of unmeasurable causal relationships, armed with theories that help them relate visible dynamic variations in systems to invisible feedback-loop structure. They attempt to guess that structure, and to include it explicitly in their models. They are searching for timeless general relationships; therefore they use data from any period and any subsystem, including, among many other sources, the same statistical data from which econometric models are derived. However, they generally prefer direct, qualitative observations of the physical processes and human actors in the real system to quantitative aggregate social indices. They would regard a series of conversations with mothers about their children to be as useful a source of information as a twenty-year time series on fertility data.

System dynamicists visualize a spectrum of increasingly precise information, ranging from intuitions, hunches, and anecdotal observations at one end to controlled physical measurement at the other, with social statistics somewhere in the middle. They declare that this spectrum offers far more information than is currently used, and that the real need is not for more data but for better use of the data already available. They point out that econometricians, by confining themselves to the narrow part of the spectrum consisting of social statistics, which contain no information about the operating policies, goals, fears, or expectations in the system, are hopelessly restricted in learning about how social systems work.

These two basic approaches to the interpretation and use of various kinds of knowledge result in continuous, fruitless cross-paradigm discussions about the relative importance of structure versus parameters. Econometricians probably spend 5 percent of their time specifying model structure and 95 percent estimating parameters. System dynamicists reverse that emphasis. Their long-term feedback models are prone to wild excursions if even one small information link is left unclosed but are often maddeningly unresponsive to parameter changes. Having worked with such models, system dynamicists find it difficult to imagine why anyone would bother to estimate most coefficients very accurately, especially when the coefficients are part of a model with an obviously defective open and linear structure. The econometrician, on the other hand, may find that in his models a 6.4 percent growth rate produces a very different result from a 7.0 percent growth rate, and his client may care a great deal about that difference. To him the system dynamicists' cavalier attitude about precise data seems both irresponsible and unsettling. Furthermore, since the econometric paradigm provides no acceptable way of finding model parameters without statistical data, he cannot imagine how a system dynamics model becomes quantified. Since the numbers are not obtained by legitimate statistical methods, they must be illegitimate, made up, suspect.

The structure-parameter split is also revealed in the complaint often voiced by econometricians that "system dynamicists deliberately design their models to generate the results they want." System dynamicists do habitually specify in advance the dynamic behavior they will regard as a first test of confidence in the model, and do operate with some knowledge about what kinds of structure will produce what kinds of behavior. However, the task of making a complex dynamic simulation model behave in any reasonable way is surprisingly difficult, especially with a closed structure, with a paradigm requiring every constant and variable to have a recognizable real-world meaning, and with a bias against including time-dependent driving functions. After working with models like that, one begins to regard the relatively sensitive econometric model as much easier to manipulate. A system dynamicist would answer the econometrician's complaint this way. "Give me an open system with five dummy variables and forty exogenous driving functions, and I *could* design a model to generate the results I want." Both complaint and countercomplaint miss the essential point—the two kinds of models are each subject to rigid constraints of different sorts and are sensitive in different ways. A scrupulous modeler in either field will feel too bound by the characteristics of the real system to engage in conscious manipulation of results, and an unscrupulous modeler in either field can get away with outrageous fiddling. Unfortunately, neither field is sufficiently self-monitored or self-critical to reward honesty or eliminate fiddling.

How Do You Know You're Right?

After each type of modeler has worked on an inherently unsolvable problem in the other's view, and has gone about it with entirely the wrong emphasis, the misunderstanding becomes complete when the finished models are examined for validity. Each kind of model fails to meet the other's criteria of validity or utility. The econometrician had a hard enough time understanding where the system dynamicist's numbers came from. Now he must evaluate the result without a single R^2 or t-test or Durbin-Watson statistic to help him along. He will find it impossible to calculate any statistical summary indices because there will be multiple covariances and colinearities and no data for many of the model's variables. The system dynamicist, who considers summary statistics either deceptive or meaningless, looks for the intuitive reality of the individual causal relations and the total dynamic behavior of the econometric model. He finds linearities, driving exogenous variables, and worst of all, dummy variables, which correspond to highest-order cheating in his paradigm. The few instances of feedback he finds will be predominately positive feedback, which he knows will carry the entire model to ridiculous extremes if forecasts are generated for more than a few years into the future.

Even sincere efforts to understand each other's evaluation techniques tend to produce classic cross-paradigm conversations such as the following dialogue (from Greenberger, 1962) between a system dynamicist and two mathematical economists, all of whom seem to be trying very hard to communicate:

Howard: We are used to seeing in the sciences one curve labeled "predicted" and another labeled "observed." These curves allow us to make evaluations such as "This is good," or "This is not so good." Is there any reason in principle . . . why you cannot take actual sales production, and inventory data, use your model to obtain predicted sales, production, and inventory figures for the corresponding period, and make a comparison?

Forrester: Yes, there is a reason why you cannot. . . . suppose you take two models, absolutely identical in structure and parameters, but both having different noise components in their decision mechanisms. If you start these models from identical initial conditions and let them run, their behaviors will diverge so quickly that there is no way of predicting what will happen on a specific day. Yet, the two models will exhibit similar qualitative performance characteristics. They will both be stable or unstable, for example. . . . Thus one must predict, not the particular event, nor the shape of the particular time history, but one must predict the change in the performance characteristics: profitability, employment stability, and characteristics such as these. The test you suggest of comparing a particular time history with the output of a model is not a test that you can expect to use, although it is a test that many people have been attempting for many economic models.

Howard: But I think that you have to have some quantitative measure of how good your model is. . . . How can we possibly criticize you when you say, "It has the same qualitative behavior"? We both look at the same simulated history, and I say it does not look at all like the real thing, and you say it does. You say that you cannot with your model duplicate the actual sales data because of the noise in the system. All you can do is get a signal that has the same characteristics as the actual data. I say that this statement has no content. . . . How can we get a quantitative agreement on what constitutes the same characteristics?

Forrester: This is a very troublesome question in the abstract, and yet in the actual specific case it is not answered in the rigorous objective sense that you speak of; neither is it in any of our real-life activities. I think you are trying for something here that we do not have in other areas of human endeavor. We do not have it in medicine or law or engineering. You are trying for something here that is more nearly perfect, more objective than in fact we know how to do anywhere else. I do not disagree with the desirability of it. I say we do not have it and we are not ready for it. Where we seem to have it in certain of the statistical model tests, I believe it is misleading and on an essentially unsound foundation.

Holt: It is interesting to contrast Professor Forrester's willingness in model formulation to quantify such unstructured concepts as "integrity" with his unwillingness in model testing to accept quantitative tests of the models. Even where *quantitative* data are available for such variables as employment fluctuations both from the company and from the model, he accepts *qualitative* judgments on similarity as perfectly adequate.

Can these two apparently antithetical ways of looking at and modeling social systems coexist within the mind of a single person? Can they coexist within the modeling profession? Or is it necessary, as Kuhn implies, that one paradigm must come to dominate the other totally?

Some people maintain that system dynamics and econometrics can indeed be merged within one person's mind and that in practice such mergers are appearing. System dynamicists who use statistical techniques to determine model parameters can certainly be found. Econometric models increasingly contain state variables, distributed lags, and feedback. But these examples are just borrowings from each other's techniques, not shifts in world view. If the problem addressed by a computer model reflects an open system view, if the model variables are observables, if the validation procedure involves detailed matching with historic data, then I would say the model is in the econometric paradigm, no matter what mathematical technique or computer language is used. If the problem is centered on generic dynamic behavior of a mostly closed system, if the variables include motivations and goals, if the validation includes assessment of the realism of the model structure, then it is a system dynamics model. I cannot imagine how the two basic *philosophies* can be mixed or merged in one model, although the tools that have shaped, and been shaped by, those philosophies might be exchanged. Perhaps, however, using the tools of a paradigm can lead to a gradual, subconscious absorption of the paradigm itself.

On the level of the modeling profession as a whole, the outcome of the econometrics-system dynamics competition may be similar to the pattern of competition between species in an ecosystem. According to the competitive exclusion principle, two species struggling for the same ecological niche cannot coexist for long. One must eventually eliminate the other completely, as Kuhn says one competing scientific paradigm eventually eliminates the other from legitimate professional practice. However, when there are *diverse niches* available, it is entirely possible for one species to lose to the other in one kind of niche and dominate the other in a different kind. Econometrics and system dynamics clearly fit different niches in the modeling policy-making environment. As long as both short-term predictions and long-term perspectives are needed, these two techniques can both be actively pursued, probably with continued mutual hostility, at least until a better competitor comes along.

2.6 Conclusion

Each modeling school defines a particular way of looking at the world and provides a set of tools for working on particular kinds of problems. None is comprehensive enough to encompass all that might be observed about the world or to solve all problems. And, of course, very many observations and problems fall far outside the range of any formal modeling method. Computer modeling can certainly contribute greatly to human comprehension and control of complex systems. But like any other tool, it must be used with wisdom and skill—and that means with understanding of its appropriate uses, of its limitations, and of the way it influences its users' perceptions of the world.

Note

1. Title suggested by a section heading in Gunnar Myrdal's *Asian Drama*. This chapter is a condensation from the forthcoming *The Electronic Oracle* by D. H. Meadows and J. M. Robinson.

References

Barr, T. N., and Hazen F. Gale. 1973. "A Quarterly Forecasting Model for the Consumer Price Index for Food." *Agricultural Economics Research,* vol. 25.

Budzik, Philip. 1975. "The Future of Dairy Farming in Vermont." Master's thesis. Thayer School, Dartmouth College, Hanover, N.H.

Cramer, J. S. 1971. *Empirical Econometrics.* Amsterdam: North Holland Publishing Company.

Forrester, Jay W. 1961. *Industrial Dynamics.* Cambridge, Mass.: MIT Press.

Forrester, Jay W. 1969. *Urban Dynamics.* Cambridge, Mass.: MIT Press.

Forrester, Jay W. 1971. *World Dynamics.* Cambridge, Mass.: Wright-Allen Press.

Greenberger, Martin, ed. 1962. *Management and the Computer of the Future.* Cambridge, Mass.: MIT Press.

Hamilton, H. R., et al. 1969. *Systems Simulation for Regional Analysis.* Cambridge, Mass.: MIT Press.

Hoffman, Robert G. 1973. "Wheat-Regional Supply Analysis." *The Wheat Situation.* WS–225. U.S. Department of Agriculture: Economic Research Service.

Hudson, Edward A., and Dale W. Jorgenson. 1974. "U.S. Economic Growth: 1973–2000." In *Long-Term Projections of the U.S. Economy.* Lexington, Mass.: Data Resources, Inc.

Intriligator, Michael D. 1972. "Econometrics and Economic Forecasting." In *Economics of Engineering and Social Systems.* Edited by J. Morley English. New York: Wiley-Interscience.

Kuhn, Thomas S. 1970. *The Structure of Scientific Revolutions.* 2d ed. Chicago: University of Chicago Press.

Maslow, Abraham. 1966. *The Psychology of Science: A Reconnaissance.* Chicago: Henry Regnery.

Meadows, D. L. 1970. *The Dynamics of Commodity Production Cycles.* Cambridge, Mass.: Wright-Allen Press.

Modigliani, Franco, Robert Rasche, and J. Phillip Cooper. 1970. "Central Bank Policy, The Money Supply, and the Short-Term Rate of Interest." *Journal of Money, Credit, and Banking,* vol. 2. 1973. Board of Governors of the Federal Reserve System. *Equations in the MIT–PENN–SSRD Econometric Model of the United States. Washington, D.C.*

Myrdal, Gunnar. 1968. *Asian Drama.* New York: Pantheon.

Naill, R. F., et al. 1975. "The Transition to Coal." *Technology Review* vol. 78.

Peck, Graham. 1967. *Two Kinds of Time.* 2d ed. Boston: Houghton Mifflin Co.

Richmond, Barry. 1976. "Conceptual Monograph No. 2." MIT System Dynamics Working Paper, Cambridge, Mass.

Schantzis, S. G., and W. W. Behrens. 1973. "Population Control Mechanisms in a Primitive Agricultural Society." In *Toward Global Equilibrium.* Edited by D. L. Meadows and D. H. Meadows. Cambridge, Mass.: Wright-Allen Press.

Schumacher, E. F. 1973. *Small is Beautiful.* New York: Harper and Row.

Spann, Robert M., and Edward W. Erickson. 1973. "Joint Costs and Separability in Oil and Gas Exploration." In *Energy Modeling.* Edited by Milton F. Searl. Washington, D.C.: Resources for the Future.

Streissler, Erich W. 1970. *Pitfalls in Econometric Forecasting.* Tonbridge, Kent: The Institute of Economic Affairs, pp. 73–74.

Tinbergen, Jan. 1937. *An Econometric Approach to Business Cycle Problems.* Paris: Herman et Cie.

Part II
Applied Principles

How Differences in Analytic Paradigms Can Lead to Differences in Policy Conclusions

3

David F. Andersen

3.1 Introduction

In each of the two cases presented here, a problem area has been studied from two different analytic perspectives. In each case, the differences in analytic paradigms have led to differences in policy conclusions. Such dependency of policy conclusions upon the mathematical perspective of the analyst has profound theoretical and practical implications. It places a responsibility upon the analyst to examine continually his selection of technique as well as his specification and execution of a study from within a given technique. Furthermore, there is a subtle interaction between the selection of a methodology and the definition of a problem. The constraints of different mathematical methodologies force the researcher to define problems precisely, so that his analysis will be tractable within the framework of the methodology chosen for the study. This accommodation of the problem to fit the methodology produces generic methodological biases whereby certain methodologies tend to "discover" policy implications well suited to, and defined in terms of, the paradigm's own a priori constraints. Different policy conclusions may be made in separate studies of the same system because the different paradigms guiding each study define problems and discover conclusions that in some sense fit within their respective frames of analysis.

Finally, the dependence of policy conclusions upon the analytic paradigm chosen should remind the analyst of the inherent inconclusiveness of mathematical analyses of social policy. No research can definitely settle a difficult policy question. A different policy study launched from a different analytic paradigm may always unearth conflicting policy conclusions. The analyst must remain aware of the inherent inconclusiveness of any one study and the continuing need to evaluate one's analytic frame of reference as well as the detailed specifications and assumptions made within a given study.

3.2 A Conceptual Framework for Understanding the Influence of Alternative Paradigms

Figure 3.1 illustrates the conceptual framework to be used in this chapter. The analyst's methodological preferences influence both the way in which difficulties become defined as problems and what methodology is chosen for a study.[1] Problem definition is not a simple and straightforward matter. Problems are defined in a circular process to fit the methods available, and the assumptions associated with a given methodology are more or less rigorously met as the problem is forced into the methodology's analytic framework. Once the problem has been defined and the methodology chosen, the ensuing analysis generates understanding of the problem area and substantive policy conclusions. The key point of interest in this study is how different methodological preferences can lead to different policy conclusions by influencing the entire problem-solving effort.

Meta-Assumptions and Specification Assumptions

Two types of assumptions—meta-assumptions and specification assumptions—are always associated with quantitative policy analyses. The distinction between these two types of assumptions can be most easily seen by examining how quantitative analyses and nonquantitative analyses are similar or dissimilar. Quantitative paradigms differ from qualitative paradigms in that the former represent social realities as mathematical expressions instead of less rigorous verbal expressions. These mathematical expressions may take on many forms, such as a closed functional form, a set of logical propositions, or a computer program. In general, some highly abstract functional form can express the generic form of a given methodology. For example, the generic least squares regression problem would be formulated as

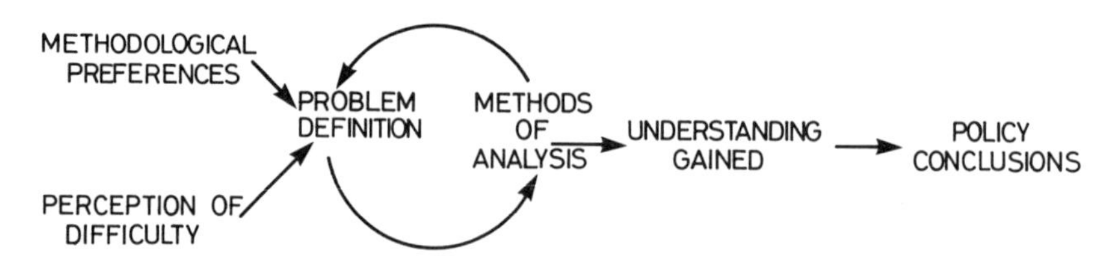

Figure 3.1

$$\underset{\text{all } \theta}{\text{Min}} \ (\hat{Y} - Y)^2$$

$$Y = F(\theta, X),$$

where one searches for the parameter vector, θ, that minimizes the squared residuals between the predicted value of Y, denoted $\hat{Y}$, and the observed Y. The predicted $\hat{Y}$ is computed as a function, F, of the parameters, θ, and the observed independent variables, X. Likewise, the generic system dynamics problem could not be formulated as

$$R = \frac{dL}{dT} = F(L, \lambda),$$

where R is a vector of rates associated with each level. These rates, in turn, are some nonlinear function, F, of the levels, L, and a vector of parameters, λ (λ may include the parameterization of table functions).

The analyst who sets out to complete a study within the framework of a given methodology knows *in advance* that his final project will conform to a certain generic form such as those sketched above. Therefore he must assume in advance that the social reality in question, or at least some significant portion of that reality, fits within the constraints of his chosen generic form.

On the other hand, the analyst has literally infinite degrees of freedom in specifying the models) for a given study. Given a reasonably robust generic form and the immense latitude of specifications offered the analyst, the analyst is all but assured that some aspect of almost any complex social situation can be treated within a given methodology.

Consequently, any quantitative study must be underwritten by two quite different forms of assumptions. Specification assumptions, the first set, are what one usually thinks about when speaking of a model's assumptions. These are evoked in a particular specification of a given generic form. They are explicitly stated and usually backed by evidence of one sort or another. They are easily reformulated and are consequently subject to manipulation and adjustment by the modeler. They are usually subjected to sensitivity testing and close public scrutiny. For example, in a system dynamics study the selection of levels, the identification of causal paths, and the formulation and parameterization of the rate equations are all specification assumptions.

The second set of assumptions are methodological priors or biases (see Randers, 1973, p. 43). These meta-assumptions are usually implicit in the generic form associated with a given methodology. Unlike specification assumptions, the analyst is not readily free to change a priori meta-assumptions (short of leaving a given methodological perspective and either inventing or adapting a new one). As such, meta-assumptions are closely connected to the generic form of a mathematical methodology and tend to dominate the world view of a given analytic paradigm.

The Role of Different Assumptions in Model Building

Meta-assumptions and specification assumptions play different roles in the problem definition stage of a study. Specification assumptions must be molded so as to best "fit" the problem statement at hand. In turn, the problem statement must be adjusted and modified so that it may be analyzed within the paradigm defined by a set of meta-assumptions. For example, within a given area of difficulty, a decision analyst will be obliged to uncover utility functions and a system dynamicist must look for closed feedback loops. However, given such constraints, the decision analyst may organize this utility measures along whatever dimensions seem most appropriate, and the system dynamicist will search for the dominant feedback effects.

The mutual accommodation between problem definition and methodological priors seems to be an inevitable characteristic of quantitative social analysis. The skilled modeler is the one who can best merge his problem definition and specification assumptions so as to capture the underlying social reality in an insightful and useful manner. Unfortunately, two skilled modelers of two different methodological persuasions may cast the same area of difficulty into subtly different problem statements, and thereby arrive at conflicting policy conclusions.

Unless one is prepared to argue conclusively that the underlying social reality conforms to a given set of meta-assumptions (for example, the wildcat oil drilling industry conforms to probabilistically branched decision trees weighted with utility functions, or simple blending problems conform to the assumptions of linear programming), or that a given practitioner is clearly more skillful, it becomes difficult to argue which analysis is best since both are probably solving different problems or focusing on different evidence.

3.3 Modeling Retirement Policies within the Enlisted U.S. Military Force

The first case considered involves two models used to analyze the effects of a shift in retirement benefits upon the total personnel costs for the enlisted personnel system of the U.S. Armed Services. Modifications to enlisted retirement benefits were being proposed by both the Department of Defense and Congress. The broad purpose of both policy models was to compute the costs and benefits caused by changes in enlisted retirement policies.

As background to the models, it should be noted that personnel may advance through a possible thirty years of service within the enlisted personnel system. During the thirty years of service, they may advance through nine enlisted grades (E–1 through E–9). Hence personnel progress through the system in two ways, both by accruing longevity (length of service) and by being promoted vertically. At any point in time, personnel may separate for one of several reasons—including voluntary separation, force-out, death, or retirement.

The first model is a static optimization model. The model was developed by an organization within the Department of Defense whose mission was to produce detailed analyses of the costs and benefits associated with various force sizes and compositions. The results of the cost-benefit studies were used to determine short-run force-management policies (such as how many personnel in a given grade and year of service should be promoted next year), as well as to provide detailed information on costs to congressional committees. This particular organization tends to view force policy questions in static and very detailed terms.

To operate the static optimizing model, the Department of Defense force analysts fed the model a host of hypothetical parameters concerning desired force characteristics. The model subsequently solved for many steady-state conditions that would produce an optimal force structure given the constraints of the hypothetical force characteristics. For example, force analysts could specify how they wanted the force to be distributed by grade, as well as what might be the lowest permissible year of promotion into each grade. The model would then compute the optimal year of service distribution and the hundreds of static promotion and advancement rates necessary to attain that optimum. It would also determine if a given set of desired force characteristics yielded no feasible optimum. The model contains an immense amount of detail and is capable of answering highly disaggregate questions. The model contains detailed costing equations and can also discount future force costs into current dollars. Policy analysts responsible for providing costs estimates to Congress for the proposed change in retirement policy perceived these detailed break-outs of costs as essential to adequately performing their missions.

However, because the model contains such immense detail, the analysts had to make several approximating assumptions to retain a tractable level of analysis. For example, promotion, retirement, and quit rate percentages had to be considered constants for the steady-state analysis. Furthermore, the model could not adequately cost out transitions from a present disequilibrium force into a hypothetical equilibrium. Under many circumstances (such as annual force-management decisions), these constraining assumptions did not seem overly restrictive, and the benefits of having detailed cost and force profile analyses outweighed any bias that these assumptions might have produced.

However, force analysts realized that the assumption of constant retention rates used in the static model might be inappropriate if retirement benefits were to be changed. Hence a system dynamics study was commissioned precisely because policy analysts suspected that significant interactions might exist between changes in force-retirement policies and the quit rates or retirement rates within the enlisted force. For example, smaller retirement benefits might make military service less attractive with respect to outside employment. Consequently, the quit rates, assumed as constants in the static model, might tend to increase. Quite predictably, the feedback emphasis of system dynamics led the second team of modelers to search for and find such effects inherent in the proposed policy revisions.

The resultant system dynamics model aggregates the thirty years-of-service cells into seven aggregate categories. The nine enlisted grades are aggregated into only three categories. This level of aggregation contrasts with the static model that explicitly models all possible categories of grade and years of service. Following their prior notions of how such a system should be treated, the system dynamicists redefined the problem as considering how changes in retirement policies might feed back to affect quit rates, retention, enlistment, and promotion rates. Again this modeling strategy contrasted with that of the static modelers who assumed constant rates in computing a steady-state optimum. Shifts in these rates bring about unanticipated shifts in force composition, both in the final equilibrium reached by the system and in the transient path into the final equilibrium. Such a redefinition of the problem, made possible by viewing the difficulty from a system dynamics approach, permitted the modeling of the interactions in a fashion not possible within the static optimization approach. Changes in retirement policies fed back to influence the current decisions of personnel to quit or to continue service.

In turn, changes in quit rates affect both the actual force profile and the number of personnel reaching retirement.

Both models showed a savings in equilibrium from the proposed changes in retirement benefits. However, the dynamic analysis led to two conclusions not found in the static model. First, the system dynamics study isolated several shifts in final force composition as potential results of implementing the proposed policy. These shifts in force composition resulted from dynamic shifts in quit rates leading more enlisted men to separate earlier in their careers thereby skewing the enlisted force toward a younger profile. Second, the system dynamics model produced a short-run cost dissavings over the first ten to fifteen years due to the transient shift in force composition immediately following the implementation of the policy. That is, higher quit rates leading to higher recruitment rates and training costs combined to form a transient increase in *total* personnel costs. However, the system dynamics model suffers as a tool for policy analysis because it could not provide the detailed cost analyses reflecting costs by individual grade and year of service usually expected from such a cost-benefit study.

A trade-off emerges between the two methodologies. The system dynamics model provides a richer description of feedback interactions because the system dynamicists define the problem in terms of feedback between aggregate system variables. However, the system dynamics model does not provide a highly disaggregate analysis of force profile and costs. On the other hand, because the first team of modelers initially chose a different form of analysis, the static model answers detailed questions concerning force profile and costs but cannot capture feedback effects. That is, the static model cannot predict how a change in retirement policies might change the final system equilibrium or affect the model's transition into equilibrium.

In sum, because of their respective differences in world view, the static optimizers were solving a detailed cost problem, whereas the system dynamicists were solving a problem centering on how quit rates change due to feedback interactions among aggregate system variables. Where one study excelled, the other fell short. Their respective policy conclusions were based upon different dynamics in the short run and different equilibria in the long run. Puzzling enough, any attempt to rank-order the studies as to which is "best" would inevitably result in each type of practitioner inventing a set of criteria flattering to his model. Then, of course, comparisons using two such sets of criteria would lead to a complex snarl of contradictions.

3.4 The Coleman Report on Equality of Educational Opportunity

The second case study, dealing with equality of educational opportunity in the United States, provides an interesting variation on the first case. The two models examined here were constructed nearly ten years apart, and the conclusions of the first study form the basis of the problem definition for the second. Taken together, these two studies of equality of educational opportunity reflect the interactive nature of quantitative social policy research, a problem solved by one analysis defining a new problem to be solved by a second generation of analysts, perhaps using a different methodology. The two studies that follow illustrate the role different paradigms and conflicting policy conclusions emanating from such paradigms can play in a dialectic development of social policy.

In 1964 Congress directed the U.S. Office of Education to

> conduct a survey and make a report to the President and Congress, within two years of the enactment of this title, concerning the lack of availability of equal educational opportunities for individuals by reason of race, color, religion, or national origin in public educational institutions at all levels in the United States, its territories, and possessions, and the District of Columbia. (Mosteller and Moynihan, 1972, p. 4, quoted from sec. 402 of the Civil Rights Act of 1964)

At that time, the study's principal author, James S. Coleman, as well as the liberal political coalition backing the law expected to find large differences in per pupil expenditures between predominantly white schools and predominantly black schools. The implicit assumption behind the Elementary and Secondary Education Acts was that a massive infusion of federal funds could reverse gross and unequal discrimination in American schools.

Two years later the report on *Equal Educational Opportunity*, or the Coleman report, was submitted to Congress. One of the most comprehensive social scientific research efforts ever undertaken, it was completed by a task force of undisputed skill and prestige. The report's conclusions were devastating to the prevailing liberal mythologies concerning the import and significance of schooling in eliminating racial inequality. Contrary to popular expectations, the report discovered only small differences in measurable educational inputs between white and nonwhite communities in the United States. Furthermore, school and teacher variables were found to have little effect in determining student achievement. These results surprised both Coleman and the political coalition that had commissioned the study.

The most controversial sections of the report were contained in its third chapter where regression techniques were used to estimate an educational production function relating educational achievement (the output) to various educational inputs such as family background, individual IQ, and aggregate indicators of social class, schooling facilities, and teaching quality. Using the regression methodology, Coleman discovered that school and teacher variables appeared to have little impact upon educational achievement.

In its original mandate, the Congress did not specify a particular approach, such as development of a production function, to the question of determining educational equality or the use of a regression methodology. Instead, Congress generally asked for an analysis of a broad difficulty. The initial dilemma facing Coleman was to define a problem that would be tractable within a recognized methodological paradigm. He settled on the problem of evaluating measurable educational inputs and outputs. Precisely put, what measurable educational policy variables best explain educational achievement? In part, such a problem definition was chosen because it fit nicely within the regression paradigm. Once he settled on the use of a regression paradigm and the "input-output" problem statement, the dilemma then centered on designing a research strategy that would violate as few of the statistical assumptions underlying the regression paradigm as possible. As illustrated in figure 3.1, Coleman was involved in the process of defining a problem to fit within his chosen analytic paradigm.

Following the distinction between classes of assumptions, once Coleman selected his approach—use of an education production function with parameters to be estimated by linear regression—he then had to assume a priori that significant policymaking inferences with respect to the national educational system could be captured within the meta-assumptions of that methodological paradigm. For example, he had to assume that major policy variables were measurable with minimal measurement error. Next, the Coleman team had to make a host of difficult methodological decisions in their treatment of specification assumptions. As samples of a few, they decided on a purely linear form for a production function and that percent of variance explained should be the measure of the impact of a variable.

When all assumptions had been made and the analysis completed, the policy conclusions were compelling and certainly counter to established intuition. The contribution of aggregate schooling and teacher indices to explaining interschool variance in student achievement was small. That is, in general, the major policy variables under the control of school officials apeared to have little or no impact on student achievement. The results were positive in the sense that strong policy implications could be inferred from the analysis (that is, many forms of direct federal aid to education would have little impact on eliminating inequality).

Both the inferences of the Coleman report and its analytic assumptions came under close scrutiny and attack. In the ten years following its publication, nearly every assumption within the Coleman report was scrutinized and several smaller replications of the study performed. Most of these critiques were launched by practitioners operating within the regression paradigm. For example, Cain and Watts attacked the use of percent of variance explained as the measure of the impact of a given policy variable. Smith examined the "effects of omitting school placement and self-selection practices on inferences about the relationships between school resources and student achievement" (Mosteller and Moynihan, 1972, p. 40). Jencks re-examined the allocation of educational resources in northern elementary schools and confirmed Coleman's assertion of small differences in educational inputs between predominantly white and nonwhite schools (Jencks, 1972, ch. 3).

All the analyses of the Coleman report began with an acceptance of Coleman's prior assumption that a production function estimated by linear regression is, in principle, capable of addressing the thorny questions associated with quality of educational opportunity. That is, studies critical of the Coleman methodology and conclusions began with the assumption that the dynamics and interactions inherent in the allocation of educational resources and the impacts of such allocations on student achievement could be addressed within the regression paradigm. The critical studies focused on the proper use of specification assumptions within the regression paradigm. The results of such reanalysis have been murky at best. Some of Coleman's conclusions were made doubtful and were partially disqualified as bases for policy analysis, but no clear-cut counterexample could be raised as long as the analysts remained within the regression paradigm.

The difficulty raised by Congress in 1964 led to the definition of a problem by Coleman (the relationship between various educational inputs and achievement) and eventually to a set of policy conclusions. Coleman's analysis and conclusions were strongly debated, but little consensus was attained with respect to their validity. Eventually, however, the debates over Coleman's results gave way to a new wave of methodological research. Researchers began to ask whether it was possible to solve Coleman's problem by means of the regression paradigm. That is, the arena of inquiry shifted to considerations of meta-assumptions.

In 1975 Luecke and McGinn replicated Coleman's analysis with an interesting twist. They used a simulation routine to generate several thousand synthetic student profiles. Four separate data-generating models were used, each of which assumed a slightly different form for the causal influence of the teacher, family, and school variables on student achievement. Although the specification of the four models differed in detail, all of the data-generating models took on a similar functional form (a Markov chain). At each point in time, each student was associated with a state vector with variables representing his family background, teacher quality, school quality, the community in which he lived (the inputs), and his accumulative achievement (the output). The achievement function either remained the same or increased by one point from one time period to the next. The probability of advancement in achievement was a function of the student's current teacher, school, home, and community variables. The exact form of this causal function varied slightly from one data-generating model to the next. Also, from one time period to the next, the student's school, teacher, or community variables could change according to a predefined causal function. These shifts represented migration from one community or school to another and the student's annual change of teachers. Each model closely followed Coleman's causal specifications except that, because they were simulation models, the modelers could keep track of the influences of individual student-teacher interactions over time as well as effects due to migration between schools that might occur over time. However, all of the specifications did assume, contrary to the Coleman report, that teacher and school variables had *strong positive effects* on student achievement.

The synthetic data profiles were then analyzed using four variations on the basic Coleman regression methodologies. The student profiles for the *last year* of the synthetic data experiment were used as cross-sectional inputs into the variations on the Coleman regression equations. Only the last year of the synthetic time-series data was used because the Coleman study had relied on cross-sectional rather than longitudinal data to make causal inferences. When the synthetic data were subjected to such regression analysis, results similar to Coleman's original results were obtained in most cases. That is, the regression analysis inferred results that were clearly contradictory to the known structural characteristics of the data-generating simulation model.

The regression model made incorrect inferences because it failed to account for the dynamics of a student's progression through school, as well as the wide variety of teachers, schools, and communities that could be experienced by a single child (and that might also have significant impacts on that child's achievement). By posing his problem in terms of a regression-estimated production function, Coleman was led to a view of the educational process and subsequently to conclusions that appear almost obviously flawed when viewed from a dynamic simulation paradigm.

Luecke and McGinn's conclusions raised doubt as to whether the regression paradigm could resolve the problems posed within the Coleman report:

> Our results suggest that studies which find little or no relationship between educational inputs and achievement may be highly misleading. *Our findings suggest that the combination of data and statistical techniques most often used is unlikely to reveal such relationships even when they exist* [italics added] . . . Researchers who conceive of education mechanistically, and use research designs which ignore the actions of individuals in schools, will find results which confirm their assumptions. (Luecke and McGinn, 1975, pp. 347–348)

By stepping outside of the regression methodology, Luecke and McGinn arrived relatively easily at a clear counterexample to Coleman's policy conclusions that years of methodological debate within the regression paradigm had failed to forcefully unearth. The mathematical form of the simulation approach allowed for a more structurally rich model that could examine detailed interaction and dynamics below the level of aggregation of the Coleman model.

In fact, Luecke and McGinn self-consciously exploited the fact that, by approaching the problem of equal educational opportunity from a different methodological perspective, they could generate a set of rigorously derived policy conclusions to counter the Coleman results. The policy inferences of the Luecke and McGinn study suggested that school and teacher variables might still matter and that policy-makers should continue to strive to improve those school and teacher variables still under their control. These policy inferences could serve as an antidote to the rather fatalistic conclusions of the Coleman study. An even stronger policy conclusion of the study was purely methodological in nature. Even if school and teacher influences were significant, studies based upon the same assumptions as the Coleman study would be unable to discover such influences.[2]

As a final note, the question of comparability between the Luecke and McGinn study and the Coleman study is puzzling. The problem attacked by Luecke and McGinn is a product of the problem solution arrived at by Coleman. Because the Coleman model is solidly based in empirical research, it could appear to be a tool for positive policy conclusions. On the other hand, Luecke and McGinn do not argue that their synthetic data represent reality. Instead, they simply used the simulation paradigm to dislodge the empirical results of the regression study.

When taken together, the two studies paint a picture of dialectic evolution in social policy. The Coleman study, based upon the regression paradigm, arrived at a compelling set of policy conclusions for American educators. Leucke and McGinn, by attacking Coleman's methodological priors, inferred a substantially different policy picture. After years of empirical and methodological controversy, the debate over equality of educational opportunity in the United States remains ill-defined and unresolved. Definition of the problem derived from Coleman's paradigmatic perspective appeared to give some resolution. But when viewed from a different perspective, both the definition of the problem and its alleged resolution appeared to weaken and lose validity.

3.5 Summary

The two cases treated here demonstrate how differences in analytic paradigms may lead to differences in policy conclusions. In the first case, the system dynamicists defined a problem focusing on feedback interactions between retirement policies and quit rates. The static modelers, guided by their prior preferences for detailed optimal solutions, defined a problem that required solving for detailed force profile and cost characteristics. Each model arrived at slightly different policy conclusions due to the differences in analytic paradigms employed.

The second case was more subtle. Luecke and McGinn used a synthetic data-simulation model to provide a counterexample to the policy conclusions of the original Coleman regression study. Conflicting policy inferences appeared as the simulation paradigm was used to launch a methodological critique of educational production functions estimated by regression.

Although section 3.2 has presented some theoretical basis for believing that such conflicts in policy conclusions *might be* inherent in any quantitative analysis of social policy, there is no apparent and easy way of knowing in advance whether or not a given set of policy conclusions is critically dependent on the particular paradigm chosen for analysis. Consequently, the two case studies suggest that a type of empirical uncertainty surrounds quantitative analyses of social policy. Each study must be veiwed as but another step in a dialectic search for social policy conclusions. The analyst must recognize both the certainty that his conclusions depend upon the methodological priors or meta-assumptions underlying his study and the probability that his conclusions may be contradicted or dislodged by a subsequent study based upon a different set of methodological priors or meta-assumptions.

Notes

1. A distinction is made between a difficulty and a problem. A difficulty is a generalized concern that draws attention to a given substantive area. A problem is a more precise specification of a difficulty. Of course, a given difficulty may lead to several interesting and fruitful problem definitions.

2. It could be argued that Coleman's faulty assumptions were specification assumptions (nonuse of time-series data) rather than meta-assumptions. However, this point is fairly moot given that nearly ten years of methodological debate *within* the regression paradigm failed to unearth the problems with dynamics and aggregation as clearly as did the Luecke and McGinn study. It might also be argued that the differences in policy conclusions stem principally from nonquantitative ideological differences between the Coleman team and Luecke and McGinn. This argument loses much credence when one realizes that the Coleman team was as surprised as anyone else over the unexpected ideological implications of their study.

References

Andersen, David F. 1975. "An Approach to the Analysis of the Military Enlisted Personnel System." System Dynamics Group working paper D–2369. Alfred P. Sloan School of Management, MIT, Cambridge, Mass.

Andersen, David F., and Robert M. Emmerichs. 1975. "Preliminary Thoughts on the Development of a Behavioral Analysis Model of the Military Personnel System." System Dynamics Group Working Paper D–2370. Alfred P. Sloan School of Management, MIT, Cambridge, Mass.

Cain, Glen, and Harold Watts. 1972. "Problems in Making Policy Inferences from the Coleman Report." *Evaluating Social Programs.* Edited by Peter Rossi and Walter Williams. New York: Seminar Press.

Coleman, James Samuel, et al. 1966. *Equality of Educational Opportunity.* U.S. Department of Health, Education, and Welfare. Washington, D.C.: Office of Education.

Jencks, Christopher. 1972. "The Coleman Report and the Conventional Wisdom." *On Equality of Educational Opportunity.* Edited by Mosteller and Moynihan. New York: Random House.

Luecke, Daniel, and Noel McGinn. 1975. "Regression Analyses and Education Production Functions: Can They Be Trusted?" *Harvard Educational Review,* vol. 44 (August 1975).

Mosteller, Frederick, and Daniel P. Moynihan. 1972. "A Path Breaking Report," *On Equality of Educational Opportunity.* Edited by Mosteller and Moynihan. New York: Random House.

Office of the Assistant Secretary of Defense (Manpower and Reserve Affairs). 1974. *Defense Enlisted Management Objectives Simulation (DEMOS) Model.* DASD, (Military Personnel Policy), Enlisted Management Systems (copy available from author).

Randers, Jørgen. 1973. "*Conceptualizing Dynamic Models of Social Systems: Lessons from a Study of Social Change.*" Ph.D. dissertation. Alfred P. Sloan School of Management, MIT, Cambridge, Mass.

The Multiplier-Accelerator Model of Business Cycles Interpreted from a System Dynamics Perspective

4

Gilbert W. Low

4.1 Introduction[1]

Paul Samuelson's famous paper (1939) on the interaction between the accelerator and multiplier was one of the first formal attempts to provide an endogenous explanation of the business cycle. The acceleration principle in Samuelson's model states that investment is proportional to the change in sales. No cycles can occur in the model without an accelerator relationship. The multiplier principle shows how a change in investment or government spending can produce over time an even greater change in total income. The multiplier itself, however, is not needed to cause cyclical fluctuations. The interaction of accelerator and multiplier produces cumulative movements in income as well as turning points.

Hicks's (1950) version of Samuelson's model has become more common than the original and is discussed in this chapter. The equations are shown as follows:

$$Y_t = C_t + I_t + G_t \tag{4.1}$$

$$C_t = cY_{t-1} \tag{4.2}$$

$$I_t = v(Y_{t-1} - Y_{t-2}), \tag{4.3}$$

where Y = income

C = consumption

I = investment

G = government expenditures (exogenous)

c,v = constants.

The model breaks time into finite periods of undefined duration. Equation (4.1) shows an accounting identity, or budget constraint, that equates income at time t with total expenditures at time t. Equation (4.2) states that current consumption is proportional to income of the past period (time $t - 1$). In Keynes's terminology, the coefficient represents a constant propensity to consume out of income, where $0 < c < 1$.

Investment in equation (4.3) follows from the assumption that firms adjust capital to a level considered appropriate with respect to total income. Time must elapse before capital can be adjusted, so (net) investment during period t is determined by the difference between desired and actual capital at the beginning of the period. Desired capital, in turn, depends on income of the previous period, which is known when decisions are made, rather than income during period t, which is unknown. Therefore, desired capital, X_t, is proportional to Y_{t-1} where the constant term v expresses the ratio of capital to output. Thus

$$I_t = X_t - K_t = vY_{t-1} - K_t. \tag{4.4}$$

Similarly,

$$I_{t-1} = X_{t-1} - K_{t-1} = vY_{t-2} - K_{t-1}. \tag{4.5}$$

So

$$vY_{t-2} = K_{t-1} + I_{t-1} = K_t. \tag{4.6}$$

Substituting, we have equation (4.3).

By combining equations, the simple multiplier-accelerator model can be expressed as a second-order difference equation for gross national product Y:

$$Y_t = (c + v)Y_{t-1} - vY_{t-2} + G. \tag{4.7}$$

Depending on parameter values, a second-order equation of this form can produce a variety of behavior modes, including oscillations (damped or explosive) and nonoscillatory movement away or toward an equilibrium.

Evans writes, "For this theory to be a workable hypothesis that can be used in conjunction with other phenomena to generate a business cycle theory, instead of just an interesting mathematical theory, reasonable values of c and v should produce damped cycles with the cycle lengths approaching those actually observed" (Evans, 1969, p. 366). The problem is that unreasonably low values of c, and perhaps of v as well, are required to produce damped cycles, and the value of time period t necessary to produce recognizable cycle lengths bears no relationship to the time in which real decisions are made.

Many revisions have been made to the Samuelson-Hicks model so as to produce more realistic behavior. A few influential examples are

Kaldor (1940), Goodwin (1951), and Allen (1960). But most of these extensions retain some version of the accelerator and underlying focus on capital investment. As Samuelson (1973, p. 260) states, "Almost all writers bring it [the acceleration principle] in as one strand in their final business-cycle theories." Moreover, in explaining the major cycles, most business-cycle theorists place crucial emphasis on fluctuations in investment or capital goods.[2]

In this chapter, I shall criticize Hicks's version of the original multiplier-accelerator model from a system dynamics perspective. The critique will raise issues that, in many cases, have been raised by other writers. But adherence to basic structural principles of system dynamics, especially the use of level variables to represent the process of accumulation, will lead to a different conclusion. In particular, I shall show that a model of the multiplier and accelerator concepts that portrays the associated accumulation of information and production flows cannot generate short-term business cycle fluctuations when reasonable parameter values are used. This finding contradicts the widely held view that multiplier-accelerator interactions underlie the business cycle.

4.2 Critique of the Multiplier-Accelerator Model

The basic multiplier-accelerator model implies important accumulations of information and physical flows but fails to show them explicitly. In this section I shall identify these accumulations, which will lead to a revised multiplier-accelerator model in section 4.3. In order to move smoothly from the critique to a reinterpretation based on system dynamics principles, I shall express the Samuelson-Hicks model in DYNAMO notation (Pugh, 1976). Figure 4.1 exhibits this formulation in a DYNAMO flow chart with the corresponding set of equations. Circles represent income, Y, consumption, C, and investment, I. Rectangles represent one- and two-period lagged values of income used in determining consumption and investment.

Lagged values of income (Y1 and Y2) are written as "information smoothing delays" (described in Mass, 1975, app. D). The equation for Y1, for example, adjust Y1 to Y over a "time to adjust Y1" TAY1. To simulate the model, time is broken into solution intervals, with a length equal to DT. When the adjustment time equals the solution interval, Y1 at time K equals the value of Y at time J (one undefined period, or solution interval, earlier). Therefore, when both adjustment times,

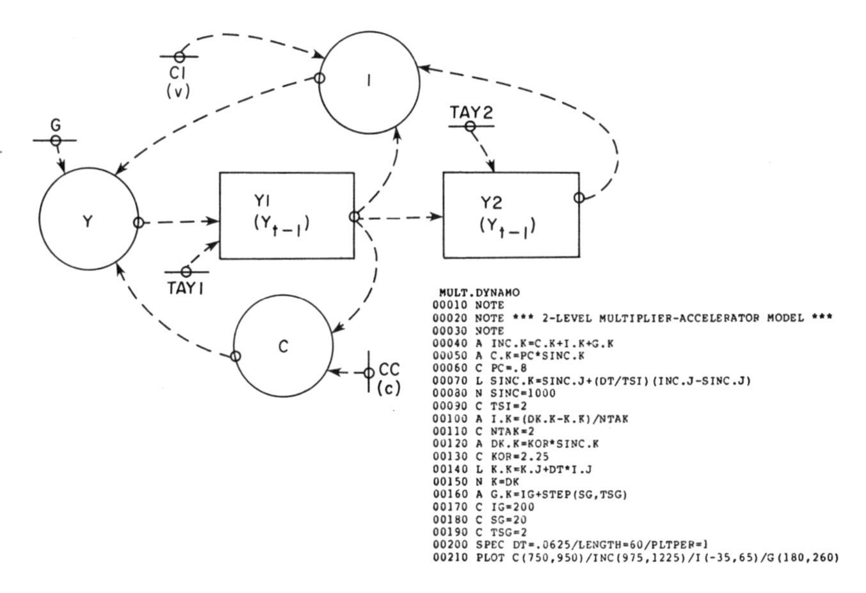


```
 MULT.DYNAMO
00010 NOTE
00020 NOTE *** 2-LEVEL MULTIPLIER-ACCELERATOR MODEL ***
00030 NOTE
00040 A INC.K=C.K+I.K+G.K
00050 A C.K=PC*SINC.K
00060 C PC=.8
00070 L SINC.K=SINC.J+(DT/TSI)(INC.J-SINC.J)
00080 N SINC=1000
00090 C TSI=2
00100 A I.K=(DK.K-K.K)/NTAK
00110 C NTAK=2
00120 A DK.K=KOR*SINC.K
00130 C KOR=2.25
00140 L K.K=K.J+DT*I.J
00150 N K=DK
00160 A G.K=IG+STEP(SG,TSG)
00170 C IG=200
00180 C SG=20
00190 C TSG=2
00200 SPEC DT=.0625/LENGTH=60/PLTPER=1
00210 PLOT C(750,950)/INC(975,1225)/I(-35,65)/G(180,260)
```

Figure 4.1
A DYNAMO translation of the Samuelson-Hicks model

TAY1 and TAY2, equal the solution interval, Hicks's difference equation model results. The model can be dislodged from equilibrium by changing the value of government spending G. A change in G will elicit convergent or divergent modes of behavior, depending on the values of CC and CI (which represent *c* and *v*, respectively, in Hicks's equations).

Accumulation of Information

To express the difference equation model in a DYNAMO format, I have used level variables to represent lagged values. But so far no distinction has been made between the solution interval and the adjustment time. That is, the level used here does not really represent the continuous accumulation of information that characterizes most decision processes. Decisions such as consumption or investment are based on accumulated knowledge, or memory, of past experience contained in the lagged values of income. Forrester (1961, p. 497) calls this accumulation "psychological smoothing" of experience, which entails the integration (in a level variable) of past information flows. Such smoothing is a gradual process of retaining recent information and then gradually forgetting the information as it recedes into the past.

The use of information delays to capture psychological smoothing processes requires a distinction between the solution interval and the smoothing delays (where DT << TAY1, TAY2). This distinction introduces a distributed lag pattern (Koyck, 1954) that differs from the previous one-period lag. Y1, for example, now equals the sum of weighted past values of Y, where the weights decline exponentially toward 0 and add up to 1. The solution interval DT is unrelated to the time delays involved in decision making but rather is required only to simulate the model through time. As DT approaches 0, in fact, a linear differential equation results. In differential equation form the multiplier-accelerator model contains four independent parameters (CC, CI, TAY1, TAY2) rather than two (CC, CI). The four-parameter model brings out the implicit information-smoothing levels buried in the solution interval t of the originial model. The time-related parameters, TAY1 and TAY2, are now explicit and subject to empirical verification.

Accumulation of Physical Flows

Introducing information levels explicitly into the investment decision brings to light a confusion in the investment equation of the original model. In both Hicks's equations and the DYNAMO translation, investment I is proportional to the difference in lagged income values (Y1 – Y2). The proportionality term, as was shown in deriving investment, represents the capital-output ratios v (CI). Dimensions in the investment equation do not match: investment is a flow of capital units per time, while the right-hand side of the equation is expressed in capital units.

This confusion reflects the assumed capacity of firms to adjust their capital stock to desired levels over one period. The one-period adjustment time is implied, but not shown explicitly, in the original investment equation. Yet in reality firms cannot always adjust capital stock when and as they wish. Depending on the availability of production capacity and finished investment goods, more or less time can elapse before plans are realized.

Available goods and capacity, however, are not represented in the model. Real investment and consumption decisions are affected not only by accumulations of information, which appear in the model, but also by physical accumulations that represent available goods and capacity to produce. By ignoring these physical accumulations, the Samuelson-Hicks model, which was designed to portray economic disequilibrium, ignores altogether the occurrence of disequilibrium in significant economic variables.

For example, Y in the model may be viewed as the sum of three components of purchases—investment, consumption, and government expenditures—or it may be interpreted as production, made possible by the receipt of three different revenue streams (the budget constraint). Production and purchases, however, are different concepts: production represents the creation of goods and services through some physical process; purchases represent the acquisition of goods once they are produced. In equilibrium, the distinction does not matter since these flows are equal. But in disequilibrium, production and purchases are not equal, and their difference must accumulate in inventories.

Inventory constitutes the physical availability of goods which can restrain purchases as inventory declines. Numerous extensions of the Samuelson-Hicks model introduce inventory as an explicit source of instability (for example, Metzler, 1941, and Abramovitz, 1950). Simpler models treat inventory as passively accumulating the difference between production and sales. More complex models adjust inventory to desired levels through changes in production but usually do not portray inventory as a source of supply than can restrain shipments in periods of low capacity and high demand. However, once a supply restraint is permitted, purchases, including the purchase of investment goods, no longer necessarily proceed at desired rates (implying changes in delivery delay). Thus fixed capital cannot always be adjusted to desired values within a prescribed period.

To account for possible supply constraints on fixed capital investment, some extensions of the Samuelson-Hicks model contain a stock adjustment principle that reflects variations in the production capacity of capital goods suppliers (for example, Kaldor, 1940, and Goodwin, 1951). In the stock adjustment formulation, investment decisions vary directly with national income and inversely with the stock of capital in existence. This formulation incorporates the basic idea of the accelerator: that investment should align fixed capital with recently prevailing income flows. A simple version of the stock adjustment principle is given by Matthews (1959, p. 41):

$$I_t = aY_{t-1} - bK_t,$$

where I_t and Y_t are investment and income, respectively, during t, K_t is fixed capital at the beginning of period t, and a and b are constants. This formulation is equivalent to the earlier accelerator equation when $a = v$ (the normal capital-output ratios) and $b = 1$ (see the derivation of equation 4.3 in section 4.1).

Capital accumulates net investment flows. As with inventories, however, this accumulation would not be important if it did not affect some other part of the system. In the capital stock adjustment models, capital only affects investment. But in reality capital also defines productive capacity and thus directly influences production—a link that is implied in these models but not always made explicit. It is the capital-production link, in fact, that defines production independently of purchases (shipments) and thus assures the possibility of imbalance between the two.

To summarize, the Samuelson-Hicks model portrays investment and consumption processes without adequately representing the accumulations of information and physical flows that underlie these processes. A level of smoothed information, implied by the model's two lagged values, enters actual purchase decisions. Moreover, the occurrence of disequilibrium in production and purchases requires the incorporation of inventory to decouple the two flows and accumulate their difference. Possible inventory shortages, in turn, mean that capital cannot always be adjusted when and as desired. Thus a model designed to produce disequilibrium behavior in income and investment should also represent capital explicitly. Variations in capital stock, in turn, will cause variations in production.

4.3 Revision of the Multiplier-Accelerator Model to Incorporate Information and Physical Accumulations

A Two-Level Model

This section revises the Samuelson-Hicks model to include accumulations identified in section 4.2. To relate the revisions directly to the original multiplier-accelerator, I shall describe a version that contains an information level (smoothed income) and a level of capital stock, but not a supply constraint in the form of inventory. Adding the inventory level follows from the critique in section 4.2 and would be a necessary step toward developing an adequate business-cycle theory. However, as shown in the second part of this section, a model with inventory would require more structure than is needed for this treatment of the early multiplier-accelerator model. The second-order model described here closely resembles Hicks's equations but produces markedly different behavior.

Figure 4.2 portrays a model that comes closest to the original multiplier-accelerator while still portraying two of the three accumulations described in section 4.2.

As in the original model, income INC represents the sum of consumption, investment, and exogenous government expenditures. Consumption C is proportional to lagged income, where in this case the lag pattern represents exponentially smoothed income SINC rather than a one-period lag. The equation for consumption (without DYNAMO time postscripts) is

C = PC*SINC,

where C consumption, PC is propensity to consume, and SINC is smoothed income.

As smoothed income SINC is an exponentially smoothed value of current income, current consumption depends on the weighted income of all past periods rather than on income of one undefined previous period. For example, with a time to smooth income TSI of 2 years, roughly 37 percent of the income level of 2 years ago is included in current "operational" or "permanent" income (and therefore influences consumption); however income of 8 years ago exerts almost no impact on today's spending. The formulation reflects the continuous process by which consumption habits and standards are gradually adapted to the levels dictated by current income and corresponds to the "permanent" portion of consumption in Friedman's (1957) permanent income

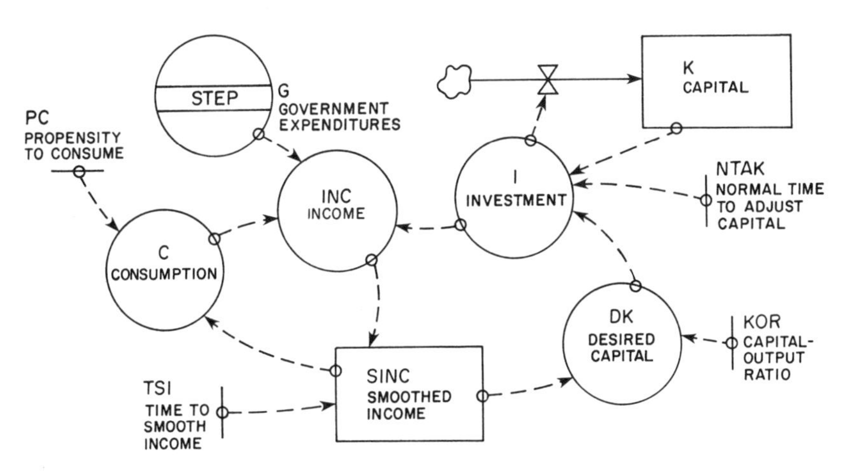

Figure 4.2
The two-level multiplier-accelerator model

hypothesis. The two-year time to smooth income TSI appears reasonable, or perhaps somewhat short. Friedman's empirical work, for example, implies a value of 2.65 for TSI (Mass, 1975, p. 85).

Consumption and smoothed income constitute parts of the multiplier loop (see figure 4.3). The direction of causality between all variables is positive, which normally produces self-sustaining, rather than self-correcting, behavior. Yet the loop is goal-seeking because the gain around the loop (which equals the propensity to consume PC) is less than 1. Suppose income is initially constant, and that government expenditures then step up in value.[3] If the loop is disconnected from the rest of the system, income and smoothed income will rise gradually to a new equilibrium over a period determined by the value of PC and the time constant used in smoothing income. The change in income after attaining a new equilibrium will be some multiple (> 1) of the step in government expenditures G. For example, given initial values specified in the model, the multiplier equals 5.

Net investment in the two-level model simply adjusts capital K to desired capital DK over a normal time to adjust capital NTAK of 2 years. The adjustment time for adding to capital reflects the period of planning and organization required to make changes in operating capacity. Empirical studies suggest a lag of about 6 quarters between appropriations and expenditures (Almon, 1965; Evans, 1969, p, 101). To this lag we should add some period for observing past activity and planning investment (Senge, 1978). The adjustment time for reducing capital would reflect the average life of capital, which would be much longer than 2 years. This asymmetry, though not reflected here, is considered later in this section.

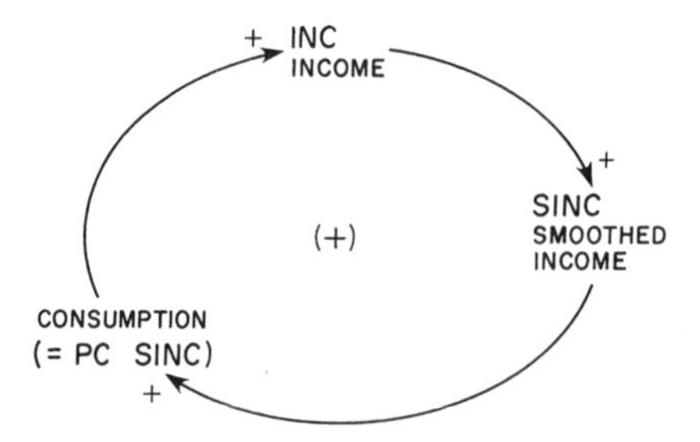

Figure 4.3
The multiplier loop

$$I = DNI = \frac{DK - K}{NTAK}.$$

Desired capital DK, in turn, is proportional to smoothed income SINC, according to the capital-output ratio KOR:

DK = KOR*SINC.

This formulation looks a lot like the basic accelerator relationship (equation 4.4). In fact, when the smoothing time for SINC is set equal to the computation interval, and when KOR = NTAK = 1, then $I_t = DK_t - K_t$ and is mathematically equivalent to Hicks's accelerator. As shown earlier, however, the mathematical equivalence obscures a dimensional inconsistency in the original investment equation and buries an information accumulation that enters actual investment decisions. Here the information smoothing relates to past activity, captured in the level of smoothed income SINC. The rationale here, as in the original model, is to adjust capital to a value dependent on lagged income. The lag pattern, however, applies exponentially declining weights to past information rather than all the weight to the previous "period."

With exponential smoothing, as well as reasonable values for KOR and NTAK, the equation is no longer mathematically equivalent to the early accelerator. But the resemblance is close, for investment is still proportional to the difference between desired and actual capital, and desired capital is still proportional to lagged income. Both of these relationships were basic to deriving the accelerator in section 4.1, (4.4) to (4.6).

Figure 4.4 identifies the accelerator loops in the revised model. Investment, like consumption, is linked to smoothed income in a positive loop and to capital stock in a negative loop. The positive loop alone is not necessarily goal seeking since the gain which depends on the value of

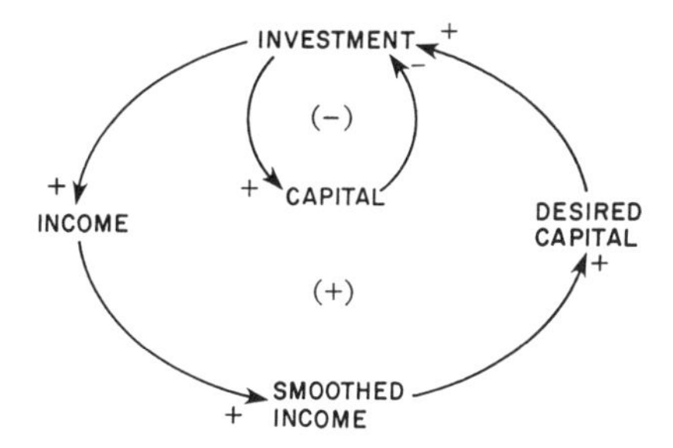

Figure 4.4
The accelerator loops

NTAK, can be greater than 1.[4] The two loops together constitute a two-level system which can produce oscillatory behavior.

In relating the revised model to actual decisions, I have suggested that 2 years is a reasonable value for the delay times (NTAK and TSI) and have distinguished these parameters from the computation interval. Combining the multiplier and accelerator loops produces the behavior shown in figure 4.5.

To generate the plot, the two-level model was disturbed form equilibrium by a 10 percent step in government purchases. Because of the step, income rises, and smoothed income (not shown in the figure) follows with a lag of 2 years. Higher smoothed income directly affects both consumption and investment. Consumption rises in proportion to the value of smoothed income and thus peaks 2 years after income. Investment also expands but peaks 2 years before income, since capital (not shown) accumulates net investment and acts through the stock adjustment formulation to prevent continued investment growth. The investment downturn eventually causes income to fall, even while consumption is still increasing. Continued growth in consumption, reflecting the operation of the multiplier, exacerbates the overshoot in income and investment and thus adds instability to the system. Net investment falls below 0 when capital stock exceeds desired levels.

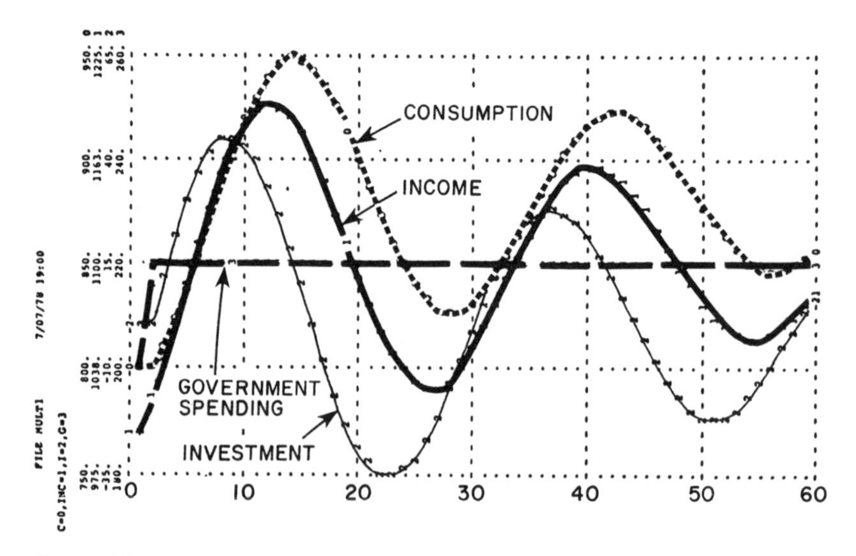

Figure 4.5
Behavior of the two-level multiplier-accelerator model in response to a 10 percent step in government expenditures

The figure displays damped oscillations with a twenty-eight-year period, a result that can also be obtained analytically.[5] Examination of the system equations shows that for virtually any set of reasonable parameter values, system oscillations, if they exist at all, will be considerably longer than the three- to seven-year period that characterizes short-term business cycles.[6] Thus with reasonable parameter values, interaction of multiplier and acclerator, in a structure that makes explicit the two levels buried in the original model, cannot produce the short-term fluctuations that the original model produces.

While the original model represents *net* investment, equal to 0 in equilibrium, a more realistic model would incorporate the asymmetric process by which capital actually grows or declines. In equilibrium *gross* investment would be sufficient to replace discards, occuring at a rate determined by the average life of capital (for example, fifteen years). In a cyclical upswing, gross investment would equal discards plus a positive capital stock adjustment term. Thus with adequate supply of investment goods, capital growth would depend on the two-year normal time to adjust capital NTAK. In a downswing, however, capital runoff would be limited by the rate of depreciation (= 1/life of capital), so that downward adjustment could take longer than upward adjustment. Tinbergen (1938), in fact, raised the issue of asymmetry in capital accumulation as early as 1938. Adding a replacement component to investment, however, does not alter our previous results, since the period of oscillation is still far too long to resemble the business cycle.[7]

Implications of Inventory Accumulation

In section 4.2 I showed that the Samuelson-Hicks model leaves out a level of inventory that in reality would accumulate the difference between production and purchase flows. In the revised two-level model just described, inventory was not included. Yet we can observe what happens to the accumulation of flows that would affect inventory by making minor changes in the revised model (see the appendix). First, an explicit production rate is implied by the constant capital-output ratio in the original model and can be added without undermining the original multiplier-accelerator concepts. Thus production would link directly to production capacity in the form of accumulated capital stock:

$$P = (1/KOR)*K.$$

Inventory then would accumulate the difference between production and the three components of income. The two-level model, revised to include passive inventory accumulation, generates behavior shown in figure 4.6. Income initially rises because of the 10 percent step in government spending, stimulating growing consumption and investment purchases. Investment accumulates in capital stock, to which production is linked. Thus production moves in phase with capital (not shown) and lags income by four years. As long as income (purchases) exceeds production, inventory in the figure falls, implausibly going below 0 by year 7.

A model containing passive inventory accumulation clearly does not resolve the criticisms raised in section 4.2, because real inventory cannot go below 0. Well before the zero point, available inventory would begin to restrain the aggregate outflow rates of consumption, investment, and government spending.

Moreover, inclusion of physical levels requires realistic control over these levels. In the case of capital stock, the control is explicit, as net investment, which accumulates in the level of capital, continuously adjusts capital to a desired value. An equivalent treatment of inventory would specify a desired inventory level and would affect the production inflow rate as well as outflow rates. Developing the model along these lines takes us considerably further from the original multiplier-accelerator concepts and is not pursued here. However, the result of

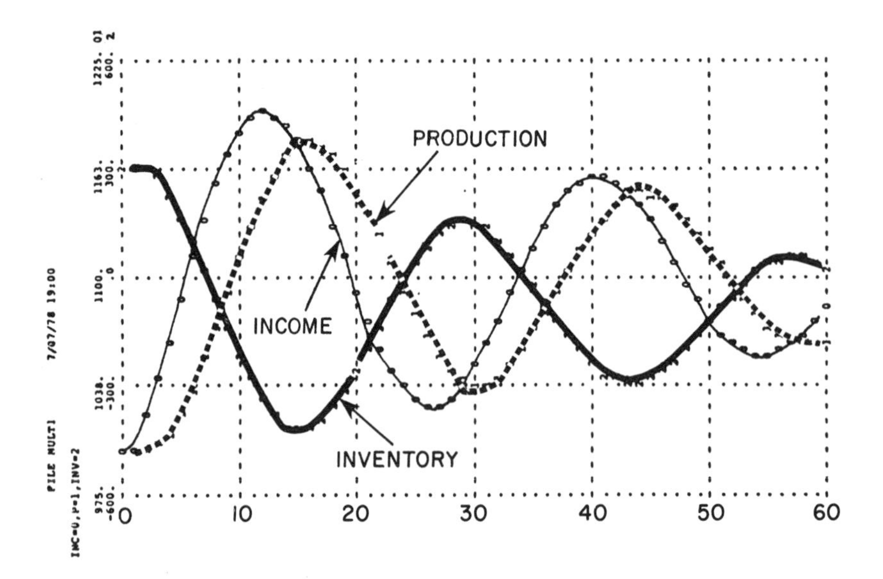

Figure 4.6
Behavior of the revised model that includes passive inventory accumulation

including a controlled level of inventory tends to confirm conclusions already suggested from simulating the two-level model (Mass, 1975).

In particular, the twenty-eight-year period brings into question the widely accepted view that managment of capital investment through the accelerator, combined with multiplier effects, underlies short-term business cycle fluctations. This long period, in fact, is more characteristic of observed fifteen- to twenty-five-year Kuznets cycles in the growth of capital and production. Mass (1975) argues in more detail that the management of fixed capital investment underlies the Kuznets cycle, just as management of capital investment in the revised multiplier-accelerator model produces a long cycle of similar, if somewhat longer, periodicity. Short-term cycles appear to reflect other influences that are not featured in investment-related explanations of the business cycle.

4.4 Conclusions

From a system dynamics perspective, accumulations and their control through explicit decision rules constitute crucial elements underlying dynamic change. Most versions of Samuelson's multiplier-accelerator model, however, omit accumulation. In this chapter, I have incorporated information and physical levels to derive a system dynamics interpretation of multiplier-accelerator interactions. This treatment has identified the levels that are implied but not explicitly portrayed in the Samuelson-Hicks model.

Accumulated information, portrayed in the revised model as an exponential average of past income, captures the "psychological smoothing" process by which memory is stored and spending decisions are made. This information level directly affects both consumption and investment demand, just as a one-period lagged value of income feeds consumption and investment in the original model. Accumulated physical flows, portrayed in levels of inventory and capital, would capture potential supply imbalances and are necessary to portray production and distribution flows in disequilibrium.

The revised model produces fluctuations with a periodicity far greater than the length of typical business cycles. This result casts doubt on explanations of the cycle, such as the multiplier-accelerator theory, that depend on the management of capital investment. Further, the analysis presented here suggests several general conclusions applicable to economic modeling:

1. Explicit focus on system levels helps to identify causal structures that underlie actual decision making. Incorporation of capital stock, for example, brings to light the direct impact of capital on production in real firms.
2. System levels, once incorporated in a dynamic model, must be controlled, just as they are in reality, by additional links between levels and rates. Adding passive inventory accumulation to the revised structure reveals the implausible behavior implied by early multiplier-accelerator models. Adding links from the inventory level to rates that control it, however, quickly moves us to a model that bears little resemblance to the basic multiplier-accelerator. In general, we cannot directly convert a model that focuses on flows to a model with accumulations that control the flows.
3. System levels clarify the direction of causality. In Hicks's equation (4.1) $Y_t = C_t + I_t + G_t$, Y may be interpreted as either production or purchases (income). When Y is interpreted as production, the equation constitutes an instantaneous budget constraint but says nothing about the causal link between production and capacity. Yet production cannot occur without capital accumulated in the past. When Y is interpreted as purchases, equation (4.1) specifies an identity only and implies no causality. Yet purchases cannot occur without available inventory accumulated over the past.
4. Portrayal of actual accumulation processes leads to more easily verifiable parameter values. In Hicks's equations, v was a parameter with uncertain dimensions or real-world meaning. Moreover, the crucial element of time was obscured by an arbitrary value, t, which confused the solution interval with the period delays. In the revised model, time constants for smoothing income and adjusting capital stock are distinguished from solution interval and constitute verifiable components of structure. Virtually any reasonable values for these parameters produce fluctuations several times longer than the short-term business cycle.
5. Models that accumulate information and physical flows often produce unexpected behavior. While many extensions of Samuelson's original model contain features similar to system dynamics versions—including inventory, capital stock, and nonlinearities—a focus on accumulation processes and their feedback effects leads to behavior considerably different from other multiplier-accelerator models. Application of system dynamics to economic analysis is continuing (Mass, 1975 and Forrester et al., 1976) and eventually may lead to deeper understanding of the business cycle and other economic processes.

Notes

1. I am indebted to several people for their helpful criticisms of earlier drafts, especially Nathaniel Mass, who worked with me on an earlier treatment of Samuelson's model, Alan Graham, Jay Forrester, John Kirsch, James Lyneis, and Peter Senge (all of MIT), and Wil Thissen, of the University of Virginia.

2. Samuelson follows Hansen in distinguishing "major" cycles of around 8 years duration from "minor" cycles of around 3.5 years and Kuznets cycles of 15–25 years.

3. The revised model retains the original assumption that government spending is exogenous. In reality, of course, rising government spending implies expanded government exployment, resulting in fewer goods being available for consumption.

4. I = (DK − K)/NTAK = (KOR*SINC − K)/NTAK.
The gain around the positive loop in figure 4.4 is KOR/NTAK which, with values shown in the model, = 2.25/2 = 1.125. Lower values of NTAK would increase the positive loop gain and thereby the system's instability.

5. The linear system is described by

$$\frac{d}{dt}\begin{bmatrix} K \\ SINC \end{bmatrix} = \begin{bmatrix} -\dfrac{1}{NTAK} & \dfrac{KOR}{NTAK} \\ -\dfrac{1}{NTAK*TSI} & \left(\dfrac{PC}{TSI} + \dfrac{KOR}{NTAK*TSI} - \dfrac{1}{TSI}\right) \end{bmatrix} * \begin{bmatrix} K \\ SINC \end{bmatrix} + \left(\frac{1}{TSI}\right) G.$$

When PC = .8, NTAK = 2, TSI = 2, and KOR = 2.25, the eigenvalue equation is

$S^2 + .0375\ S + .05 = 0$,

which yields

$S_{1,2} = -.01875 \pm i*.22282$.

6. The table that follows shows periodicities for nine combinations of NTAK and TSI, based on solutions of the eigenvalue equations. An asterisk indicates no oscillations.

		NTAK (years)		
		1	2	3
TSI (years)	1	*	26.8	26.8
	2	19.9	28.2	34.4
	3	31.2	*	*

7. In the two-level model without discards, initial equilibrium values were Y = 1000, G = 200, C = 800, implying propensity to consume PC = .8. In a model with discards, investment would be positive in equilibrium, implying that PC ≠ .8 for G = 200. If we assume average life of capital ALK = 15, then equilibrium I = discards = K/15, where K = KOR*1000 = 2250. Thus initial equilibrium values would be Y = 1000, G = 200, I = 150, C = 650, implying PC = .65. With this value for PC and the original values for other parameters, eigenvalues for a model with discards are

$$S_{1,2} = -.05625 \pm i*.217,$$

which yields a period of 28.9 years.

References

Abramovitz, M. 1950. *Inventories and Business Cycles.* New York: National Bureau of Economic Research.

Allen, R. G. D. 1960. "The Structure of Macro-Economic Models." *Economic Journal* (March), pp. 38–51.

Almon, S. 1965. "The Distributed Lag between Capital Appropriations and Expenditures." *Econometrica,* vol. 33, pp. 178–196.

Evans, M. K. 1969. *Macroeconomic Activity, Theory, Forecasting, and Control.* New York: Harper and Row.

Forrester, J. W. 1961. *Industrial Dynamics.* Cambridge, Mass.: MIT Press.

Forrester, J. W., Mass, N. J., and Ryan, C. J. 1976. "The System Dynamics National Model: Understanding Socio-Economic Behavior and Policy Alternatives." *Technological Forecasting and Social Change,* vol. 9, pp. 51–68.

Friedman, M. 1957. *A Theory of the Consumption Function.* Princeton: Princeton University Press.

Goodwin, R. M. 1951. "The Nonlinear Accelerator and the Persistence of Business Cycles." *Econometrica,* vol. 19, (January), pp. 1–17.

Hicks, J. R. 1950. *A Contribution to the Theory of the Trade Cycle.* London: Clarendon Press.

Kaldor, N. 1940. "A Model of the Trade Cycle." *Economic Journal,* vol. 50 (March), pp. 78–92.

Koyck, L. M. 1954. *Distributed Lags and Investment Analysis.* Amsterdam: North-Holland Publishing Co.

Mass, N. J. 1975. *Economic Cycles: An Analysis of Underlying Causes.* Cambridge: Wright-Allen Press.

Matthews, R. C. O. 1959. *The Trade Cycle.* Cambridge, England: Cambridge University Press, p. 41.

Metzler, L. A. 1941. "The Nature and Stability of Inventory Cycles." *Review of Economic Statistics,* vol. 23, (August), pp. 113–129.

Pugh, J. 1976. *DYNAMO User's Manual.* 5th ed. Cambridge, Mass.: MIT Press.

Samuelson, P. A. 1939. "Interactions between the Multiplier Analysis and the Principle of Acceleration." *Review of Economic Statistics,* vol. 21 (May), pp. 75–79.

Samuelson, P. A. 1973. *Economics.* New York: McGraw-Hill.

Senge, P. M. 1978. *"The System Dynamics National Model Investment Function: A Comparison to the Neoclassical Investment Function."* Ph.D. thesis, MIT, Cambridge, Mass.

Tinbergen, J. 1938. "Statistical Evidence on the Acceleration Principle," *Economica* (May), pp. 164–176.

Appendix: DYNAMO Equation Listing of Revised Multiplier-Accelerator Model Used to Produce Figure 4.6

```
 MULT1.DYNAMO
00010 NOTE
00020 NOTE *** 2-LEVEL MULTIPLIER-ACCELERATOR MODEL ***
00030 NOTE
00040 A INC.K=C.K+I.K+G.K
00050 A C.K=PC*SINC.K
00060 C PC=.8
00070 L SINC.K=SINC.J+(DT/TSI)(INC.J-SINC.J)
00080 N SINC=1000
00090 C TSI=2
00100 A I.K=(DK.K-K.K)/NTAK
00110 C NTAK=2
00120 A DK.K=KOR*SINC.K
00130 C KOR=2.25
00140 L K.K=K.J+(DT)(I.J)
00150 N K=DK
00160 A G.K=IG+STEP(SG,TSG)
00170 C IG=200
00180 C SG=20
00190 C TSG=2
00200 NOTE
00210 NOTE *** INV INACTIVE IN 2-LEVEL MODEL,
               BUT USED FOR FIGURE 6. ***
00220 NOTE
00230 L INV.K=INV.J+(DT)(P.JK-C.J-I.J-G.J)
00240 N INV=300
00250 NOTE
00260 NOTE *** PRODUCTION FORMULATED ONLY
               TO PRODUCE FIGURE 6. ***
00270 NOTE
00280 R P.KL=K.K/KOR
00290 NOTE
00300 NOTE *** RERUN INSTRUCTIONS ***
00310 NOTE
00320 SPEC DT=.0625/LENGTH=60/PLTPER=1
00330 PLOT C(750,950)/INC(975,1225)/I(-35,65)/G(180,260)
00340 RUN FIGURE 5
00350 PLOT INC,P(975,1225)/INV
00360 RUN FIGURE 6
```

Stock and Flow Variables and the Dynamics of Supply and Demand

5

Nathaniel J. Mass

5.1 Introduction

Supply and demand are the two central concepts in both classical and modern economic analysis. Models of economic processes can be divided into two broad categories: equilibrium theories, which analyze the outcome of market transactions once supply and demand for a particular commodity have settled into balance in equilibrium, and disequilibrium theories, which treat the behavior of the economy when supply and demand are not necessarily equal at every point in time.

In both equilibrium and disequilibrium models, the question arises of how to represent basic supply-demand forces that affect production rates, price movements, wage negotiations, and other economic changes. This chapter contrasts two particular viewpoints for analyzing supply and demand. The first viewpoint dominates economic thinking. It treats supply and demand as rates of flow. For example, John Maynard Keynes's *General Theory* popularized the concept of aggregate demand, which is the sum of planned consumption, investment, and government expenditures. These three forms of expenditures are all rates of flow, measured in goods units (or dollars) per unit time. Even long predating Keynes, however, the static theory of the firm regarded supply as a flow of production determined by the equalization of price and marginal cost. Analogously, the theory of the household treated demand as a flow of consumption governed by relative prices and marginal utilities.

An alternative perspective to the rate-of-flow viewpoint concerning supply and demand sees supply and demand primarily as stock variables or integrations. According to the stock-variable viewpoint, supply, for example, would be measured by the available inventory of a commodity while demand would be measured by a backlog of unfilled orders.

The distinction between stocks and flows is well known to economists. Yet economic theories still revolve primarily around flow concepts of supply and demand. An important reason for emphasis on rates of flow in economics is that both the theory of the firm and the theory of the household evolved out of a set of equilibrium concepts of profit and utility maximization, respectively. The theory of the firm determines the equilibrium rate of production that yields a maximum flow of profits; analogously, the theory of the household determines the equilibrium rates of purchase that maximize the flow of utility from current purchases. In equilibrium, inventories held by firms and households are at their desired levels. Consequently, there are no inventory discrepancies to generate upward or downward pressure on rates of production and transaction. As a result inventories and other stock variables do not typically appear in equilibrium models. As Kenneth Boulding has noted,

> In fact the theory of the firm, and of the economic organism in general, has . . . developed . . . along the lines of static equilibrium theory of "maximizing behavior." The concept of the balance sheet, unfortunately, has not been employed to any extent in developing the static theory of the firm, so that as generally presented in the textbooks the firm is a strange bloodless creature without a balance sheet, without any visible capital structure, without debts, and engaged apparently in the simultaneous purchase of inputs and sale of outputs at constant rates.[1]

The concept of equilibrium predominates economic analysis. Large numbers of models in the economics literature derive equilibrium prices and quantities based upon an assumed equality of supply and demand. In such models, supply and demand are both usually considered as rates of flow, measuring equilibrium rates of activity. But equilibrium concepts do not just underlie strictly equilibrium theories. Because of the strong theoretical basis of economic analysis in equilibrium theory, dynamic models in economics have also tended to concentrate on relationships between rates of flow to the exclusion of stock concepts. For example, Paul Samuelson's classic multiplier-accelerator model interrelates rates of production, consumption, investment, and income flow;[2] the model does not explicitly include inventories, capital stock, money levels, or other stock variables that intervene between rates of flow. Analogously, Kenneth Arrow and Marc Nerlove present a model in which price changes are governed by excess demand for a commodity.[3] In the model, they assume that excess demand (essentially the net of consumption less production) is a function of the prices and expected prices of the commodity and (potentially) all substitute and complementary commodities. Their model thus allows for the possibility of disequilibrium between production and consumption. But the model does not account for changes in inventories, backlogs, and other stock variables that would occur in a disequilibrium mode, nor for the way in which changing stock variables feed back to influence supply, demand, and prices.

This chapter attempts to demonstrate that stock-variable concepts of supply and demand must be incorporated explicitly in economic models in order to capture the rich disequilibrium behavior characteristic of real socioeconomic systems. More specifically, a number of broad implications for economic theory and modeling practice are raised:

1. Stock equilibrium and flow equilibrium, as conventionally defined in economics, will generally not arise concomitantly. Thus, stock variables will frequently be out of equilibrium, thereby causing continuing change in rates of flow, even once flow equilibrium between production and consumption has been attained. Consequently, dynamic behavior and stability characteristics of an economic system can be analyzed only by comprehensively interrelating stock and flow variables.
2. Economic systems are characterized by interactions between stock and flow variables that yield complex adjustment paths to equilibrium. Such systems cannot a priori be considered stable in the sense that an initial disequilibrium will be countered within a very short lapse of time.
3. Efficacy of policies designed to influence such disequilibrium economic behavior as economic growth or instability can be assessed properly only in a model that interrelates stocks and flows.

Several of the above points have already been recognized by economists. For example, Duncan Foley and Miguel Sidrauski (1971) discuss the need to incorporate both stock and flow equilibria in a macroeconomic growth model. Nonetheless, this chapter contributes to economic theory in three major respects: first, by synthesizing in a single discussion the diverse functions of stock variables and the motivations for including them in economic models; second, by highlighting the inherent theoretical deficiencies of equilibrium models and the large number of dynamic economic models that do not incorporate explicitly all stock variables connecting rates of flow; and third, by suggesting a concrete direction for refinement and extension of economic theory and model-building practice.

5.2 Influence of Stock Variables on Dynamic Economic Behavior

Stock variables play a major role in the dynamics of supply and demand. This section presents seven points, each describing a particular function or mechanism through which stock variables give rise to dynamic disequilibrium behavior. The list of functions is not fully inclusive but seems to encompass the most relevant points from the standpoint of economic theory.

Stock Variables Link Rates of Flow

As noted earlier, economic models of the firm center around production, consumption, and prices. In a real firm, stocks of in-process goods and final output intervene between the processes of production and consumption. If production exceeds consumption, inventory will accumulate. Conversely, if production is less than consumption, inventory will be drawn down. Typically, a firm will try to maintain inventory levels that are proportional to its internal level of activity, measured by the firm's average rate of production or sales. Discrepancies between actual and desired inventory generate pressures to expand or contract production by acquiring or disposing of factor inputs. Therefore existence of inventories links production and consumption rates, allowing them to be unequal at each point in time, while typical inventory-management policies tend to equalize production and consumption over a longer run. Boulding (1950) discusses in an analogous manner how the firm acts to preserve a "homeostasis" of its asset structure (its stocks of physical and financial assets). Production and consumption need not be equal at each point in time. But if production exceeds consumption, inventories will rise above desired levels, thereby signalling a need to contract output. Conversely, production below consumption will cause falling inventories and pressures to expand output. Consequently, imbalances between production and consumption can be corrected through the physical mechanism of production changes motivated by inventory shortages or excesses, as well as by price changes. Such quantity adjustments to a market imbalance can exert important effects on disequilibrium behavior. For example, Axel Leijonhufvud notes that the "revolutionary element in Keynes's *General Theory* was the reversal of the Marshallian assumption of infinitely fast price adjustments relative to quantity adjustments."[4]

Stock variables can link production and consumption through several other mechanisms besides inventory changes. For example, suppose that the order rate for a firm's output exceeds production capacity of the firm.

Order backlogs will tend to rise, thereby lengthening the firm's delivery delay (the average period of time required to fill an order). High delivery delay, in turn, can depress incoming orders through lack of availability of the product. Consequently, whereas price is regarded in economic theory as the fundamental market-clearing mechanism, both availability and price in fact serve jointly as market-equilibrating channels.

Finally, it should be noted that price changes in a firm arising from supply-demand pressures tend to be based on the relative magnitude of stock, rather than flow, variables. For example, upward price pressure may reflect low inventories (indicated inadequate supply) or high order backlogs (indicated excess demand). In either instance, excess demand would raise price and thereby induce expansion of production, leading to a build-up of inventories and reduction of order backlogs. By contrast to the relative size of inventories and backlogs, the balance of production and consumption in a firm does not provide a reliable indication of excess demand or supply. Suppose that production exceeds consumption. Does this discrepancy necessarily indicate excess supply in the market? It might if, for example, inventories exceed desired levels as a consequence of high production. But, alternatively, high production rate could be a *consequence* of high desired production due to low inventories or high order backlogs. In this instance, production in excess of consumption would be a consequence of *excess product demand.*

More attention should be given in economic theory to the way in which stock variables such as inventories and backlogs trigger price and quantity adjustments. Robert Clower and Axel Leijonhufvud remark that

> A theory capable of describing system behavior as a temporal process, in or out of equilibrium, requires a prior account of how trade is organized in the system. Equilibrium, steady-state theory has managed pretty well without such an account. Macroeconomic theory cannot do so. Microeconomic theories of how business and household units behave—of how production and consumption decisions are made—when the system is not in equilibrium will have to be predicated on some such account.[5]

By expanding their theories to encompass the stock variables linking rates of flow, economists will necessarily move in the direction, advocated by Clower and Leijonhufvud, of filling in the details of "how trade is organized" in a complex production-consumption-distribution system.

Stock Variables Can Induce Opposing Short-Term and Long-Term Effects

Classical economics asserts that an increase in marginal costs of producing a commodity should lower supply, by shifting the supply curve so that a lower supply is elicted by a given price; lower supply, in turn, should drive up prices due to excess demand. However, real economic systems display a much more complicated pattern of interaction among price, supply, and demand. One frequently observed pattern of behavior exhibits opposing short-term and long-term price responses.

To take a concrete example, opposing short-term and long-term price changes are exhibited by the behavior of hog prices in commodity markets during 1971. In 1971 corn prices rose dramatically due to a severe corn blight in the midwestern states. Increased price raised the marginal costs of hog production, since corn is the primary feed for hogs. Many economists expected increased costs to lead to higher prices and lower supply of hogs. But in fact, hog prices declined in 1971 and increased back to their 1971 level only about a year later, subsequently continuing to rise. From the standpoint of static equilibrium analysis, such a pattern of price behavior appears anomalous, but the behavior becomes readily explicable through expanding the traditional flow-oriented concepts of supply and demand to encompass the level variables (stocks) in a typical commodity system.

Figure 5.1 shows the essential stocks and flows that characterize a commodity system such as hog farming. Live hogs are held in two forms: in a mature stock where they are fed for approximately two months before slaughter; and in a breeding stock where hogs are withheld from market for breeding purposes. The breeding stock determines the breed-

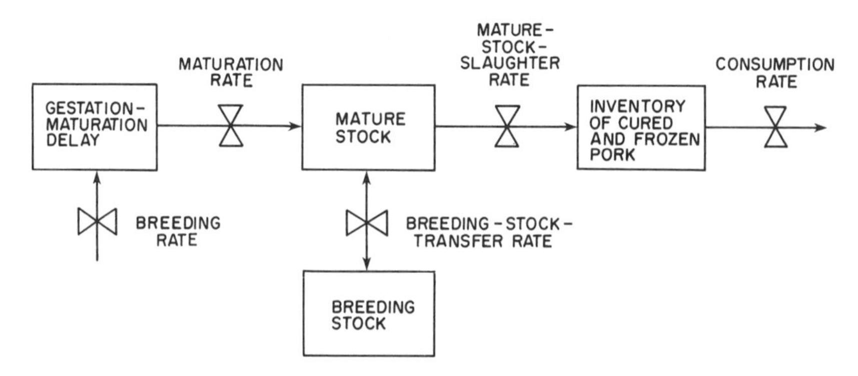

Figure 5.1
Stocks and flows in a typical commodity system

ing rate of hogs. Young hogs flow into the mature stock category after a ten-month gestation-maturation delay.

Of significance in figure 5.1, no single variable can alone be termed the "supply of hogs." Instead, at least four distinct concepts are related to the supply of hogs. First, the size of the breeding stock determines the breeding rate, which is equal to the maturation rate in equilibrium. Second, the size of the mature stock measures the number of hogs soon to be coming on the market. Third, the mature stock slaughter rate measure the annual rate of addition to final output inventories of cured and frozen pork (analogous to final production rate). Fourth, the inventory of pork measures the amount of the product available for immediate sale and subsequent consumption. As will be shown later, the different variables that measure the supply of hogs can be moving in opposite directions at the same time, thereby generating opposing short-run and long-run pressures on price.

What will be the response of the system shown in figure 5.1 to an increase in marginal costs of production? Rising marginal costs lower the profitability of raising hogs and consequently induce producers to lower desired breeding stock. Producers therefore tend to transfer hogs from breeding stock to mature stock to reduce breeding stock to its desired level. As a result the short-term response to increased marginal cost is to reduce breeding stock, expand mature stock, and increase the flow of hogs onto the market (figure 5.2). Thus *expanded short-term supply and downward price pressure result from increased marginal costs.* The increased supply and lowered price are not due to irrational responses on the part of producers but are largely necessary *physical consequences* of the structure of the commodity system. Over the longer term as breeding stock is reduced, breeding and maturation rates will decline. Final output inventory will thereby decline, generating upward pressures on price.[6] This long-term effect is the outcome anticipated in the classical economic analysis of supply and demand, where an increase in marginal costs leads to an equilibrium of higher price and lower production rate. However, the classical result encompasses only the long-term response to supply, while the short-term response runs in the opposite direction.

Two points are illustrated by the hog-production example. First, in an economic system, supply and demand may have multiple manifestations, some being expressed through stock variables. Second, supply and demand can each change in opposing short-term and long-term directions. Analysis of these divergent short-run and long-run impacts re-

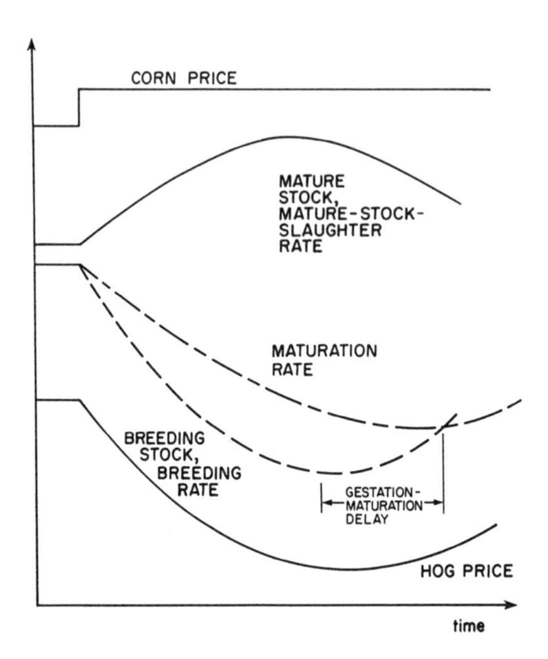

Figure 5.2
Opposite short- and long-term adjustments in a commodity system

quires explicit portrayal of the stock variables that influence supply and demand.

Stock Variables Amplify Rates of Flow through Accumulation Effects

Stock variables can induce amplification of rates of flow such as production and consumption. The term "amplification" refers to the tendency for a response in an economic system to exceed the amount of change that would at first seem to be entailed by the causes of that response. An example of amplification would be a 10 percent increase in production rate induced by a 5 percent increase in incoming order rate.

To demonstrate one source of such amplification, consider the response of production to a step increase in consumption (shown in figure 5.3 as a step increase in incoming orders). Higher consumption depletes inventories and subsequently induces expansion of production. Production eventually intersects consumption at point t_1. However, all through the period from t_0 to t_1 inventory has been steadily depleted. Therefore at point t_1, inventory would be approximately at its minimum value. Concurrently desired inventory would have risen in response to

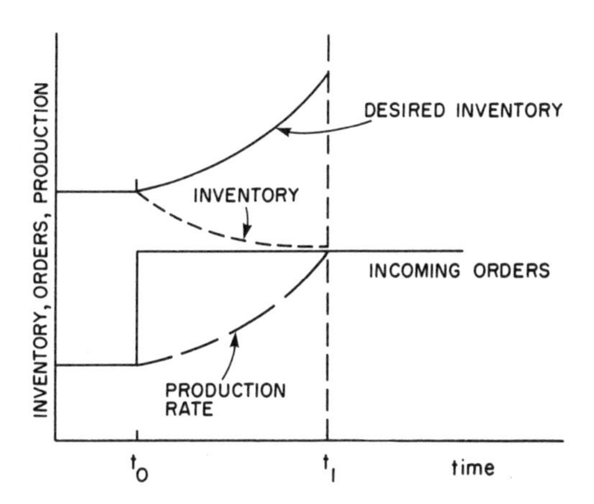

Figure 5.3
Production overshoot caused by inventory-management policies

increased consumption (sales). The resulting inventory imbalance impels continued expansion of production above consumption. Production must expand above consumption in order to rebuild the inventories depleted while production was still below consumption, and to build inventories up to a higher absolute level set by the increased desired inventory. Therefore even when supply and demand are equal in the rate-of-flow sense, supply-demand pressures embodied in stock variables can move the system out of its flow equilibrium. In this simple production-consumption example, rebuilding the diminished inventory necessitates an increase in production above consumption. Resulting amplification of production can easily cause successive over- and underexpansion of produciton relative to consumption. Such amplification effects cannot be captured in economic models that are confined to interrelating rates of flow.

Viewing supply and demand as stocks or level variables begins to show how economic processes frequently considered to be stabilizing mechanisms in fact may be destablilizing or at least can prolong disequilibrium behavior. Looking back to figure 5.3, suppose that prices tend to rise as long as inventories are adequate, indicating insufficient product supply. Then prices would be high and rising most rapidly at the point where production equals consumption, thereby encouraging a production overshoot. Consequently, in a real economy containing inventories, backlogs, and other stock variables, prices may well have a net *destabilizing*, rather than *stabilizing*, effect on economic activity. In other words, production behavior can be less stable when production policy responds to price than when production is governed solely by physical mechanisms of inventory and backlog correction. Such issues merit careful further investigation. Economic models must realistically treat the full range of mechanisms governing disequilibrium behavior to yield proper conclusions about stability of economic systems and desirability of alternative economic stabilization policies.

Stock Variables Underlie Multiple Modes of Economic Behavior

In an economic system, different time constants or speeds of adjustment may be associated with different stock variables. In turn, differences in adjustment times may give rise to multiple modes of economic behavior. Analysis of such multiple modes is important from a theoretical and policy standpoint. To the extent that separate processes underlie different modes of behavior, different points of intervention and different policy levers may be called for to influence each mode.

To show how multiple behavior modes can arise, consider an economy containing two factors of production—labor and fixed capital stock. Labor is augmented through hiring and decreased through separation. Fixed capital stock is increased through investment and decreased through depreciation. Labor and fixed capital differ in two important respects. First, in an economy such as the United States, labor can be acquired fairly readily over a period of weeks or, at most, months. By constrast, construction and delivery of fixed capital require a longer period, perhaps one to three years. Acquisition of new capital equipment must also frequently be preceded by a long planning period. During that time technical specifications are drawn up, plans are debated and modified, appropriations are approved, and credit is negotiated if the project is to be financed through debt or equity issues as distinguished from internal finance. Thus labor has a short time constant or delay in its planning and acquisition, compared with fixed capital.

Labor and fixed capital also differ in their turnover times. In the United States, labor can be discharged on very short notice or reduced fairly quickly through attrition. Average duration of employment is approximately two years.[7] By constrast to labor, fixed capital is a durable asset with an average lifetime of ten to twenty years or longer. Thus labor has a much shorter time constant for turnover than fixed capital.

To see the behavioral significance of the different time constants associated with labor and fixed capital. Consider a production process in which both labor and fixed capital contribute to production. Behavioral impacts of adjustments in labor and fixed capital can be isolated first by holding one factor input constant, subjecting the system to a change in incoming orders, and examining the resultant behavior; then by holding the other factor input constant and repeating the same analysis; finally by allowing both factors to vary and studying the resulting behavior.

To apply the above framework, start by assuming that fixed capital stock is constant, so that production rate can be altered only through changing the level of labor or through short-term changes in capacity utilization. If incoming orders increase, overshoot of production and employment will occur in accordance with the inventory and backlog correction mechanisms described earlier. That is, production must expand above consumption, and employment must increase in the short term over its eventual equilibrium value, to replenish depleted inventory stocks and build inventory up to a higher level necessitated by an increased level of business activity. As described in detail by Mass (1975, 1976), such interactions between employment and inventories can produce fluctuations in employment, inventory, and production characteristic of short-term business cycles in the economy.

Now reverse these factor-input assumptions, and suppose that labor is held constant while fixed capital stock is allowed to vary. If incoming orders now increase, overshoot and fluctuation in fixed capital stock and production will tend to occur as a consequence of the same structural mechanisms described above. Production must overshoot consumption to build up inventories, whether the underlying factor of production is labor or fixed capital. In other words, the mechanisms producing overshoot and fluctuation are structurally parallel for the two cases of variable labor input and variable capital input.

The difference in behavior in the two instances described above will be primarily in the periodicities of fluctuation. That is, cycles associated with adjustments in fixed capital will be relatively long compared with labor-adjustment cycles. As described by Mass (1975), time constants of adjustment associated with labor and fixed capital may differ sufficiently so that, when labor and fixed capital are both allowed to vary, the economy exhibits a short-term cycle due to labor adjustments superimposed on a longer-term cycle caused by fixed capital-investment policies. The results imply that labor hiring and firing policies primarily govern the short-term business cycle and that fixed capital investment is chiefly involved in generating economic cycles of much longer duration, contrary to the prevalent capital-investment theories of business cycles.[8]

From a policy standpoint, the results suggest importance of evaluating economic stabilization policies according to their impacts on both short-term and longer-term cycles. Such policy evaluation must be performed using models that treat explicitly the various stock variables and time constants that underlie multiple modes of fluctuating economic behavior.

Stock Variables Can Propagate Long-Term Economic Changes

Looking at processes of supply and demand in terms of stock variables provides insight into mechanisms through which long-term changes can be propagated through the economy. As just demonstrated, the process of fixed capital investment may underlie economic cycles of longer duration than the short-term business cycle. For example, Mass (1975, 1976) shows how capital accumulation processes can underlie the fifteen- to twenty-year Kuznets cycle of growth in capital plant and potential output. Going further, Forrester (1975), and Mass and Forrester (1976) suggest that interactions between fixed-capital-producing and capital-consuming sectors may be involved in generating fifty-year long waves in the economy. Such long-run behavioral phenomena arise from the way in which disequilibrium values of stock variables within a system promulgate disequilibrium rates of flow; such flows, in turn, cause long-term changes in stock variables characterized by relatively long time constants of adjustment. The particular significance of the capital-investment example cited above lies in the fact that fixed capital investment has traditionally been considered an essential factor in generating the short-term business cycle. However, consideration of the accumulation processes governing fixed capital leads to the conclusion that process of investment are too slow to interact appreciably in a cycle of only a few years duration. The position that capital investment is principally involved in generating economic cycles of much longer duration than the short-term business cycle has previously been argued by Moses Abramovitz and others, but remains a minority viewpoint among economists.[9]

Stock variables may also capture attitudinal factors that influence long-term economic development. For example, a recent article dealing with the reasons for overbuilding of office space in New York City describes how long-term attitudes toward risk affect successive building cycles.[10] Office space in New York was significantly overexpanded in the 1920s. The resulting severe financial losses on the part of developers and financial institutions led to an introduction of stringent lending standards. In particular, developers planning to construct a new office building had to have the office space 75 percent leased out before they could obtain long-term credit. Such policies guaranteed at least 75 percent occupancy in office buildings, thereby alleviating the threat of overexpansion. However, as time passed, lending standards were gradually relaxed as recollections and fears of the overexpansion in the 1920s subsided and as individuals responsible for introducing the original standards either retired or passed away. The culmination of these declining standards was another massive wave of office construction, leading to high office building vacancy rates in the 1970s.

Attitudes toward financial risk represent a part of the state or condition of the socioeconomic situation and change slowly in response to economic and social forces. As such, they can be described as stock variables or integrations of past attitudes and circumstances. Increased recognition is needed in economics of how socioeconomic forces embodied in stock variables, including social values and attitudes, and the underlying process of integration can generate long-term disequilibrium behavior.

Stock Variables Measure the Determinants of Economic Welfare

The classical theory of the household assumes that households maximize utility subject to the budget constraint that the value of purchases not exceed present income. Household utility, or welfare, is in turn assumed to be a function of current rates of purchase of the various goods and services. There are several flaws in this static framework. First, the budget constraint specifying that income must equal the total value of household purchases is an equilibrium condition where money balances held by households remain constant. In disequilibrium, however, income may exceed purchases, leading to net accumulation of money balances; or, alternatively, income may fall short of purchases if consumers draw down their existing money balances. In either instance, changes in money levels lead to changes in ability or willingness of consumers to spend money in the future. For example, consumer spending in excess of present income will deplete money balances, thereby lowering future spending power.

A second defect of the classical theory of the household is described by Boulding:

> The error in question is the identification of income, either in the form of production or consumption, with economic welfare, or perhaps it would

be more accurate to say the use of income as a measure of economic welfare. So ingrained is this identification in our thinking that the assumption passes almost unquestioned, not only in the economics of the neoclassical school as represented by Pigou, but also in the more fashionable Keynesian economics. . . .

The illusion that consumption—and its correlative, income—is desirable probably stems from too great preoccupation with what Knight calls "one-use goods," such as food and fuel, where the utilisation and consumption of the good are tightly bound together in a single act or event. We shall return to the problem of one-use goods later. In the meantime, let us direct our attention toward many-use goods, such as houses, automobiles, furniture, crockery, clothing, machinery and tools, buildings, roads, bridges, etc. It is quite clear that the consumption of these goods (which necesitates their production) is something quite incidental to their use and frequently not even closely connected with the degree of use. We want houses, not because they depreciate, get dirty, sag, crack, disintegrate, and need repairs; we want houses because we can live in them, and the living in them is in no way bound up with their consumption. If we had houses that would not depreciate, walls that would not get dirty, or require painting, roofs that would never leak, foundations that would never sag, furniture that would not wear out, crockery that would not break, footwear that never needed repair, clothing that never got ragged or unpressed, we would clearly be much better off: we would be enjoying the services of these things without the necessity of consuming or producing them. Coming now nearer to the one-use goods, consider fuel—that the consumption of fuel for domestic heating merely arises because of the depreciation of warmth by poor insulation; and economy in the consumption of fuel that enables us to maintain warmth or to generate power with lessened consumption again leaves us better off. . . .

There are important implications of the above analysis, both for economic theory itself and for the policy conclusions which stem therefrom. In the first place it is necessary to separate more clearly than hitherto the concept of income, output, or gross national product from the concept of economic welfare. There may be, and usually is, a correlation between the level of income and or welfare. But this connection is by no means invariable, and it would be most rash to suppose that an increase in income always means an increase in welfare.[11]

According to Boulding then, consumer welfare depends on available stocks of goods and services, rather than on the rates of addition to, or subtraction from, these stocks. Recognition that acts of consumption involve accumulation processes provides insight into consumption impacts on the economy. First, changes in price or income which affect consumers' desired stocks can exert accelerator-type changes on production, employment, and income. For example, if price of a particular commodity rises, consumers' desired stocks of that commodity will decline. Purchase rate of the commodity must then fall steeply below usage rate in the short run in order to allow consumer-held inventories of the commodity to deplete. However, as the level of the commodity declines, even if price remains constant, purchase rate will begin to increase toward the commodity usage rate. Second, purchases of durable goods are deferable since the utility provided by the good depends on the available stock rather than on the rate of purchase. Deferability of purchases depending on the ratio of desired to actual stocks of goods can induce consumption cycles where purchases alternately exceed and fall below the equilibrium rate of purchase and usage.

Modern consumption theory has tended to depart from the assumption of utility maximization subject to the budget constraint that the value of present consumption not exceed present income. For example, Franco Modigliani's "life-cycle model" of consumption assumes that a household plans its consumption over its entire lifetime "to redistribute the income it gets (and expects to get) over its life cycle in order to secure the most desirable pattern of consumption over life."[12] However, consumption theory still does not adequately incorporate the diverse stock-variable influences on consumption. For example, utility is still assumed to be derived from the purchase rate of goods and services rather than from available stocks. Moreover, consumption functions in economic models are seldom accompanied by explicit internal accounting for houshold money pools and stocks of consumption goods, and infrequently consider the feedback which these stock variables exert on purchase rates. Consequently, such phenomena as deferrability of durable purchases are frequently overlooked even in modern consumption functions. To summarize, while consumption theory appears to be moving away from short-term equilibrium analysis, much further refinement is needed to capture all the relevant stock-variable effects that influence consumer welfare and consumption behavior.

Stock Variables Produce Variable Delays that Induce Overshoot and Oscillation

Changes in stock variables can produce variable delays and stock-flow ratios that contribute to disequilibrium behavior.[13] For example, if demand for a particular commodity rises, rising order backlogs and declining output inventories of the commodity can lengthen delivery delay for the commodity. In turn, as delivery delay rises, consumers of the product will order further ahead or order more than they really need to be able to maintain their desired usage rate.

The mechanisms through which varying delivery delays can produce fluctuations in consumption were described in a 1924 article by Thomas W. Mitchell. He hypothesized an initial situation in which retailers, caught short of inventories, increase their orders for goods. As goods are shipped, manufacturers' inventories are depleted, thereby creating shortages and raising delivery delay for goods. At this point, according to Mitchell,

> . . . [r]etailers find that there is a shortage of merchandise at their sources of supply. Manufacturers inform them that it is with regret that they are able to fill their orders only to the extent of 80 percent; there has been an unaccountable shortage of materials that has prevented them from producing to their full capacity. They hope to be able to give full service next season, by which time, no doubt, these unexplainable conditions will have been remedied. However, retailers, having been disappointed in deliveries and lost 20 per cent or more of their possible profits thereby, are not going to be caught that way again. During the season they have tried with little success to obtain supplies from other sources. But next season, if they want 90 units of an article, they order 100, so as to be sure, each, of getting the 90 in the pro rata share delivered. Probably they are disappointed a second time. Hence they increase the margins of their orders over what they desire, in order that their pro rata shares shall be for each the full 100 per cent that he really wants. Furthermore, to make double sure, each merchant spreads his orders over more sources of supply.
>
> Herein originates a large false demand upon manufacturers, and herein lies a great defect of our system of competitive private initiative in industry. . . . [T]he false demand is passed back, stage by stage, along the channels of production. . . . What, in turn, is the natural result of this situation? Eventually the streams of production are not only enlarged but overenlarged. There comes a time when the ultimate sources of supply fill nearly all the orders of their customers. The latter are surprised to find their orders filled promptly and fully, and that they are receiving more than a plentiful supply of materials. There is no longer a shortage. Instead, owing to their previous overordering, there is a surplus. Their rate of ordering slows up a little, and the ultimate sources of supply find business not quite so brisk. The producers in the second stages also fill their orders promptly and fully, thus surprising their customers in turn. Result, orders upon the second stages in the production process slow up a little. And so on down to the retailers. The rivers of production have swollen so that the volume of flow is no longer insufficient to fill the apparent capacity of the market as evidenced in orders. Indeed, pro-

duction has come to exceed the real demand, and the capacity of production organizations. . . .[14]

More attention should be given in economics to analyzing how variable delays produced by changes in system levels influence short-term and long-term disequilibrium dynamics. Such analysis requires explicit representation of both stocks and flows, and consideration of the dynamic changes that can occur in the ratios of system levels to rates of flow through them.

5.3 Conclusions

This chapter has delineated some of the major mechanisms through which stock-variable concepts of supply and demand affect short-term and long-term economic behavior. Because of its historical foundation in equilibrium analysis, economic theory has tended to revolve around relationships between rates of flow, such as production and consumption. Wassily Leontief has noted that "exclusion of stocks from the original input-output scheme limits its applicability as a general equilibrium theory to short-run analysis."[15] But stock variables can cause significant disequilibrium changes through accumulation effects and varying stock-flow ratios, even over periods as short as several months (see section 5.2). Analysis of stability characteristics and dynamic properties of an economic system therefore requires comprehensive treatment of stock variables that intervene between rates of flow.[16] Expanding economic analysis in this direction should enhance our capabilities for understanding economic dynamics and make headway against the policy problems of managing a disequilibrium economy.

Notes

1. Boulding, 1950, p. 34.

2. Samuelson, 1939.

3. Arrow and Nerlove, 1958.

4. Leijonhufvud, 1970, ch.2; also see Clower, 1965.

5. Clower and Leijonhufvud, 1975, p. 183.

6. Figure 5.2 shows that increased price will eventually reverse the decline in breeding stock, thereby leading to an increase in breeding and maturation rates and in the inventory of cured and frozen pork.

7. *Statistical Abstract of the United States*, 1970, p. 218.

8. See Samuelson, 1939, Duesenberry, 1958, Hicks, 1950, and Kaldor, 1940, for a description of major capital-investment theories of the business cycle.

9. See Abramovitz, 1961.

10. Eleanore Carruth, 1975, "The Skyscraping Losses in Manhattan Office Buildings," *Fortune* (February).

11. Boulding, 1949–1950, pp. 77, 80, 83.

12. Modigliani, 1957, p. 105. See also Ando and Modigliani, 1963.

13. By contrast, for example, dynamic input-output analysis *assumes* that stock-flow ratios are always constant over time (Leontief et al., 1953, ch. 2). By virtue of this assumption, such analysis misses a whole range of disequilibrium phenomena.

14. Mitchell, 1924, pp. 645–647.

15. Leontief et al., 1953, p. 12.

16. Econometric models sometimes try to capture processes of accumulation (integration) implicitly, through distributed lag formulations directly connecting rates of flow. However, such practice appears generally undesirable or infeasible. First, use of distributed lag formulations tends to obscure underlying accumulation processes and thereby move the model away from observable description of real life. Second, and more important from a theoretical standpoint, time constants and delay times across stock variables will seldom be constant. For example, delivery delays for goods will depend on suppliers' available stocks of those goods, and the turnover time of labor in the economy will depend on the multiple time-varying factors that influence termination rates and voluntary quit rates (see Runge, 1976 for a comprehensive model of labor flows in a multisector economy). Such variable time constants are easily neglected in a distributed lag formulation that statistically relates an "input" and "output" without a clear conception of the real-life processes through which the two are linked. Or at best, trying to incorporate variable time constants in a distributed lag frequently requires nonlinear formulations that can pose formidable statistical problems in estimation. For these reasons the most sound practice, both theoretically and empirically, is to formulate models to explicitly include all relevant stock variables.

References

Abramovitz, Moses. 1961. "The Nature and Significance of Kuznets Cycles." *Economic Development and Cultural Change*, vol. 9 (April), pp. 225–248.

Ando, Albert, and Franco Modigliani. 1963. "The 'Life-Cycle' Hypothesis of Saving: Aggregate Implications and Tests." *American Economic Review*, vol. 53 (March), pp. 55–84.

Arrow, Kenneth J., and Marc Nerlove. 1958. "A Note on Expectations and Stability." *Econometrica*, vol. 26 (April), pp. 297–305.

Boulding, Kenneth E. 1945. "Consumption Economics: The Consumption Concept in Economic Theory." *American Economic Review Papers and Proceedings*, vol. 35 (May), pp. 1–14.

Boulding, Kenneth E. 1949–1950. "Income or Welfare." *Review of Economic Studies*, vol. 42, pp. 77–86.

Boulding, Kenneth E. 1950. *A Reconstruction of Economics*. New York: John Wiley and Sons, Inc.

Boulding, Kenneth E. 1973. "The Economics of the Coming Spaceship Earth." In *Toward a Steady-State Economy*. Edited by Herman Daly. San Francisco: W. H. Freeman.

Clower, Robert W. 1965. "The Keynesian Counterrevolution: A Theoretical Appraisal." In *The Theory of Interest Rates*. Edited by F. H. Hahn and F. P. R. Brechline. London: St. Martin's Press.

Clower, Robert W., and Axel Leijonhufvud. 1975. "The Coordination of Economic Activities: A Keynesian Perspective." *American Economic Review*, vol. 65 (May), pp. 182–188.

Duesenberry, James S. 1958. *Business Cycles and Economic Growth*. New York: McGraw-Hill.

Foley, Duncan K., and Miguel Sidrauski. 1971. *Monetary and Fiscal Policy in a Growing Economy*. New York: Macmillian Co.

Forrester, Jay W. 1976. "Business Structure, Economic Cycles, and National Policy." System Dynamics Group Working Paper D–2245–2, MIT, Cambridge, Mass. Also published in *Business Economics*, vol. 11 (January 1976), pp. 13–24.

Hicks, John R. 1950. *A Contribution to the Theory of the Trade Cycle*. London: Clarendon Press.

Kaldor, Nicholas. 1940. "A Model of the Trade Cycle." *Economic Journal*, vol. 50 (March), pp. 78–92.

Leijonhufvud, Axel. 1970. *On Keynesian Economics and the Economics of Keynes*. New York: Oxford University Press.

Leontief, Wassily, et al. 1953. *Studies in the Structure of the American Economy*. New York: Oxford University Press.

Mass, Nathaniel J. 1975. *Economic Cycles: An Analysis of Underlying Causes*. Cambridge, Mass: Wright-Allen Press.

Mass, Nathaniel J. 1976. "Modeling Cycles in the National Economy," *Technology Review*, vol. 78 (March–April), pp. 2–12.

Mass, Nathaniel J., and Jay W. Forrester. 1976. "Understanding the Changing Basis for Economic Growth in the United States." System Dynamics Group Working Paper D–2392–2 MIT, Cambridge, Mass., Also published in *US Economic Growth 1976–1986:* "Forecasts of Long-Run Economic Growth." *Prospects, Problems, and Patterns,* series vol. 4. Congress Joint Economic Committee (1976).

Mitchell, Thomas W. 1924. "Competitive Illusion as a Cause of Business Cycles." *Quarterly Journal of Economics*, vol. 38 (August), pp. 631–652.

Modigliani, Franco. 1957. "Savings Behavior: A Symposium." *Bulletin of the Oxford Institute of Statistics*, vol. 19 (May), pp. 99–124.

Runge, Dale. 1976. "Labor-Market Dynamics: An Analysis of Mobility and Wages." Ph.D. dissertation, MIT, Cambridge, Mass. (June).

Samuelson, Paul A. 1939. "Interaction Between the Multiplier and the Principle of Acceleration." *Review of Economics and Statistics*, vol. 21 (May), pp. 75–78.

Part III
Conceptualization

Guidelines for Model Conceptualization

6

Jørgen Randers

6.1 Introduction

In spite of the existence of innumerable social system models, there is not much available literature, and probably not much existing knowledge about the process by which such models are constructed. How is a problem chosen? How does one achieve a useful perspective on the problem area? How does one succeed in capturing the essentials of a complex, real-world phenomenon in a relatively simple model. Models are, nearly without exception, presented in final form, as though such a thing as The Model exists and as though the process of arriving at a fruitful description of reality is straightforward and not worthy of explicit attention. The lack of information about the modeling process, particularly its first stages, is probably due to the "pre-scientific" state of the art of modeling.[1] Model conceptualization is especially difficult in the modeling of social phenomena because social systems are more complex and less well understood than physical systems, and because the modeler must represent aspects of the real-world that are not easily observed or measured.

Because there is no educational text on model conceptualization, the sequence of presentation in published papers describing models are commonly mistaken for the actual steps in the creation of those models.

It is only through personal experience that modelers gradually develop effective procedures for model construction, and discover that these do not resemble the picture conveyed by an effective presentation of the final result.

In this chapter I describe one effective procedure for model conceptualization, give two examples, and highlight a number of common mistakes made by the novice. This procedure can be applied to a wide range of modeling efforts. Perhaps reading about it will help the novice develop better working habits in less time than if he were to discover the procedure on his own. Unfortunately, few beginners resist the temptation to follow what is felt to be the "natural" approach to modeling: a headlong rush into description of the real world in the form of flow diagrams. It is only after several unstructured and unsuccessful attempts that the intense desire for guidelines evolves. Before this, the beginning modeler rarely appreciates the immense strength and utility of the approach to model conceptualization that is recommended here. Thus the following may seem more useful to the experienced modeler than to the novice.

6.2 The Process of Modeling

The modeling process can be divided into four stages: conceptualization, formulation, testing, and implementation. Even before they are described, it must be stressed that the four stages do not follow each other in tidy sequence, neither in practice nor ideally.

The conceptualization stage establishes the focus of the study—the general perspective and the time horizon. The critical decisions are made on what part of reality to study and how to describe it. The formulation stage casts the chosen perspective into a formal representation. The resulting model gives a precise, though not necessarily accurate, description of a slice of reality and is capable of generating images of alternative futures. The testing stage subjects both model structure and behavior to various tests intended to establish the quality of the model. The goal is to identify weak points for further improvement and to establish the extent of model utility. Finally, the representation stage seeks to transfer study insights to those that might use them. The full process is listed in table 6.1. Only a few points will be made here concerning formulation, testing, and implementation. These topics are treated in detail elsewhere in this volume.

It is worth spending a few words on the repetitive nature of the modeling process. As most other forms of product development, modeling is a process of trial and error. The path toward a useful model typically looks like the jittery curve illustrated to the right in figure 6.1. The goal of an effective procedure for model construction is not to remove *all* iterations but to achieve a reasonably consistent degree of progress throughout the recursive process.

The system dynamics paradigm and the DYNAMO computer language can be seen as tools to facilitate the conceptualization and

Table 6.1
The four stages of model construction

Conceptualization	Familiarization with the general problem area
	Definition of the question to be addressed—either, What caused a given development? or, What are the likely effects of a given policy?
	Description of the time development of interest (the reference mode)—defining the time horizon and the range of time constants in the model
	Verbal description of the feedback loops that are assumed to have caused the reference mode (the basic mechanisms) —defining the system boundary and the level of aggregation
	Development of powerful organizing concepts
	Description of the basic mechanisms in causal diagram form
Formulation	Postulation of detailed structure—selecting levels, selecting rates and describing their determinants
	Selection of parameter values
Testing	Testing of the dynamic hypothesis—Do the basic mechanisms actually create the reference mode?
	Testing of model assumptions—Does the model include the important variables? Are the assumed relationships reasonable? Are parameter values plausible?
Implementation	Testing of model behavior and sensitivity to perturbations
	Testing the response to different policies
	Identification of potential users
	Translation of study insights to an accessible form
	Diffusion of study insights

formulation stages, respectively. In the complex, unstructured conceptualization stage the modeler strives toward a "mental model," that is, an understanding of the operation of the real world. The mental model is a prerequisite for formulating any formal model. The system dynamics theory of system structure and behavior is a powerful aid in the conceptualization process, for instance, in drawing the attention of the analyst toward closed loops of cause and effect and in stressing the distinction between levels and rates. In the subsequent much simpler formulation stage, the modeler writes down his mental model in a formal, accessible description. The DYNAMO language is a conducive medium for the formulation task—according to the system dynamics paradigm,

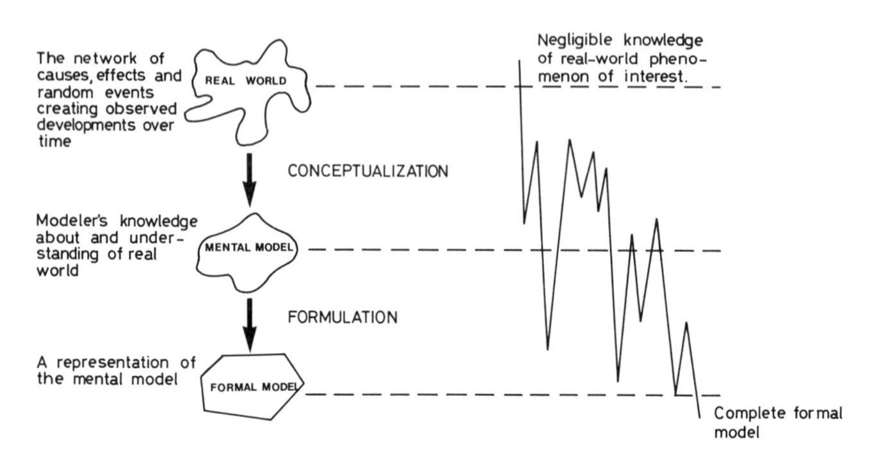

Figure 6.1
The recursive nature of the conceptualization and formulation stages

facilitating "good" modeling and placing nearly insurmountable barriers against "bad" formulations.

6.3 The Social Process as a Basis for Model Conceptualization

When discussing the process of model conceptualization, it is absolutely necessary to distinguish between a social *system* and a social *process*—or in system dynamics terminology, between structure and behavior.

A social system is a set of cause and effect relationships. A model structure is a map of the slice of the real world chosen as the "system" under study. The set of rules and administrative practices that are established to control inflation, along with the channels through which decisions are transferred to real-world change and then into observations, is an example of a social system.

A social process is a chain of events—that is, a time development. The simulation output from a system dynamics model is meant to portray such a sequence of events in the real world. The stop-and-go character of economic development may serve as an example of a social process.

In the early phases of model construction, it is far more helpful to view a system dynamics model as a model of a social process than as a model of a social system. Experience shows that in order to establish a useful model it is more productive to grasp hold of a social process and ask about its cause than it is to select a slice of the real world and ask what behavior it will generate.

This point is worth stressing because there is a strong "natural" tendency to start by describing the system. The tendency may arise from the tangible character of the system as compared to the elusive nature of its time development, and from the fact that model structure is normally presented before model behavior in final reports.

Of course, the ultimate goal in system dynamics modeling is to understand more about the links *between* structure and behavior, in other words, about the structural origins of a given time development and about the behavioral consequences of changes in the system. So in the final end the modeler is equally interested in structure and behavior. But not so in the early stages of model construction. Here it proves advantageous to start with identification and description of the time development of interest, and only then proceed to identification and description of the underlying causes.

To facilitate the discussion, I define two terms to denote the social process and the social system chosen for study. The time development of interest will be called "the reference mode." The reference mode serves as a tangible manifestation of the entity that is being portrayed by the model output.The smallest set of realistic cause and effect relations that is capable of generating the reference mode will be called "the basic mechanisms." The basic mechanisms are the central elements in one's understanding of the causes behind the time development of interest.

6.4 The Reference Mode as a Guide to Model Structure

Unless one knows exactly what social process the model is meant to portray, it is very difficult to decide on what to include or to exclude from the model structure. Since all cause and effect relationships have *some* effect on real-world developments, there is a strong temptation to include any causal mechanisms that come to mind. The result is a complex cluttered structure lacking in focus. While it is true that a model can be obtained even in this manner, it is uncertain whether this model will run in a plausible way, and it is almost certain that it will not give insight.

The reference mode is intended to solve the problem of specifying the study focus. What is a reference mode? It is a graphical or verbal description of the social process of interest. The reference mode of a model under development can be stated by drawing a graph of the expected behavior of major variables. Doing so helps to define more clearly which variables must appear in the model. The reference mode may also be stated by a discussion of the phenomena the model is meant to portray. A verbal treatment of the reference mode may actually convey more of the purpose of the model than a graph of expected behavior. Often the reference mode encompasses different possible time paths for the model variables.

The identification and description of a reference mode greatly facilitates the selection of the basic causal structure of the model. First of all, it simplifies decisions on what to leave out: any mechanisms that is not believed to be a major cause behind the reference mode should be left out of the initial model. But the reference mode also keeps the attention of the modeler squarely fixed on what matters, namely, the variables involved in the mode and possible explanations for their observed change over time. And finally, the use of reference mode as a first standard for model behavior ensures that the initial model embraces the essential dynamics. In short, the modeler is guaranteed that the model runs are reasonable and interesting.[2]

We are here, however, in a situation where general advice and guidelines lack concrete meaning to the unexperienced modeler. Practical examples of reference modes, their use as the basis for selecting the basic structure and as the benchmark for the behavior of the initial model, is worth more than repeated exhortations. Thus I turn to two examples.

Case 1: A Historically Observed Reference Mode

The simplest case arises when the modeler in his effort to familiarize himself with the problem area stumbles across a historical time series that illustrates and concretizes the problem at hand in terms of a dynamic development. This time series can be used as a reference mode. Once a reference mode is chosen, it provides solid foundation for the next stop in the modeling process—which would be to create an initial model that is capable of reproducing the major dynamics of the reference mode. The reference mode helps keep the initial model simple and transparent; in other words, it restricts the model to the basic mechanisms. Such transparency makes the model easier to understand and simpler to explain.

In a recent modeling study, John Høsteland was interested in the recurring problems of low profitability in the Norwegian pulp industry.[3] In the early phases of the study it was tempting to go straight ahead and try to gather in one model all the commonly suggested causes of low profitability. This approach fit well with the advice from experienced people in the industry, most of whom believed in different causes for each wave of low profitability, and with the criticism from fellow academics who could be silenced only by including their pet relationships in the model. As could be expected, this initial approach led to large confusing models generating uninteresting runs. In fact the runs could be described as straighforward "sums" of the mechanisms included in the model, and as such gave little *new* insight.

Then Høsteland came across time-series data which when (smoothed and) plotted looked as shown in figure 6.2. Conversations with industry people that constantly returned to the problems of excess inventories, and the typical system dynamics fascination with inventory fluctuations made Høsteland choose figure 6.2 as his reference mode. The whole study now became oriented toward answering the following question: What are the basic causes behind figure 6.2 ? Which proved to be a well-defined task compared to the fumbling exploration in the earlier attempts to gather all causes of low profitability in one structure.

Of course, there was no clean-cut agreement about the causes of the observed four-year cycle in pulp inventories. But figure 6.2 served as a productive and concrete starting point for discussions both with industry

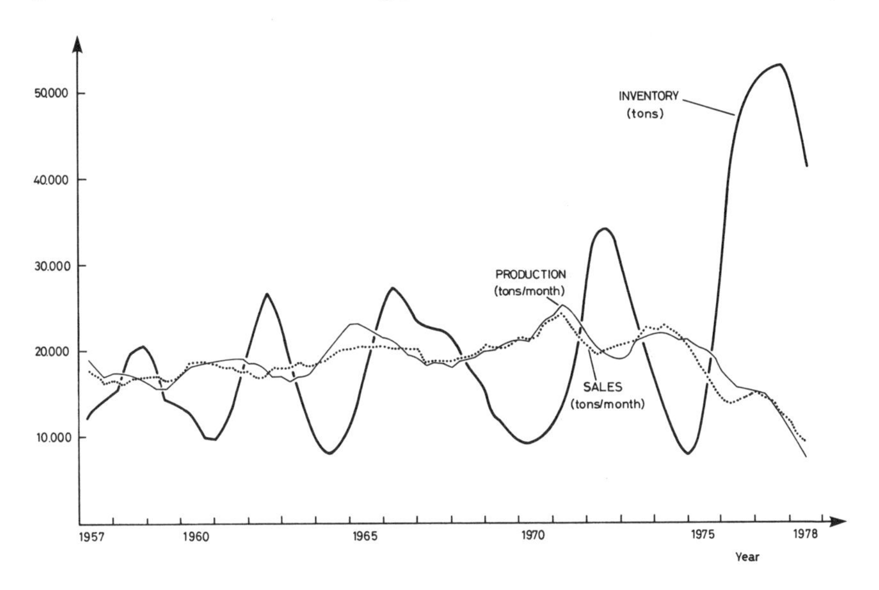

Figure 6.2
The reference mode for a study of recurring problems of low profitability in the pulp industry showing cycles in inventory, production, and sales

people and other analysts. Before long the causal loops in figure 6.3 were established as the most likely basic mechanism behind the cycle—in addition to the oscillation imposed by the four-year business cycle in the international markets where the Norwegian pulp is sold.[4] The basic mechanisms were embedded in an initial model, which proved to exhibit a slightly damped cycle with the right period and amplitude, even without exogenous driving forces.

It is easy to proceed from this basis, that is, from a simple model recreating the essential dynamics of an observed time development of interest. One will typically strive toward better reproduction of the historical behavior, or toward a more credible model for evaluation of alternative policies. In both cases the task will be to improve and often to detail the description of the basic mechanisms.

Case 2: A Hypothesized Reference Mode

The procedure described in case 1 becomes less clean-cut in a situation where it is impossible to find a historical illustration of the reference mode, because either the mode has not yet occurred in reality or data are not available.

The same procedure can be used, however. But in this case the study must be based on a *hypothesized* reference mode. This introduces one difficulty: the modeler can no longer be certain that the reference mode is "correct." He must be much more willing to reshape the reference mode based on insight gained in the early modeling stages. The reference mode and the initial model thus interact in a bootstrap pattern whereby both are gradually improved with the help of the other.

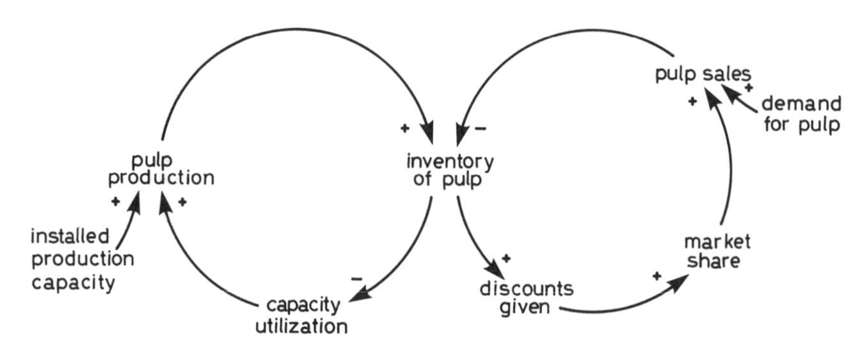

Figure 6.3
The basic mechanisms thought to underlie the inventory fluctuation in figure 6.2

The development of the World 3 model of growth in a finite world serves as an example of the use of a hypothesized reference mode.[5] Familiarization with the general problem of material growth in a physically finite world was acquired through professional discussion, news reports, and the literature. After much thought, the question to be addressed was defined: Will human material activity adjust smoothly to the global carrying capacity or go through a period of overshoot and collapse? Further exploration led the modelers to emphasize the erodability of the carrying capacity (for instance, through soil erosion due to intensive agriculture, or destruction of the self-cleansing capacity of the ecosystem due to excessive pollution). Sudden reductions in the earth's ability to sustain its population seemed possible due to over-utilization.

The possibility of collapse because of excessive load placed on the physical environment drew attention to the development sketched in figure 6.4a. The behavior mode of overshoot and decline appeared to be a possible consequence of current trends. Since overshoot was judged undesirable, it seemed worthwhile to investigate possible causes of that behavior and try to determine how a change in growth policies might achieve a more gradual adjustment like that depicted in figure 6.4b.

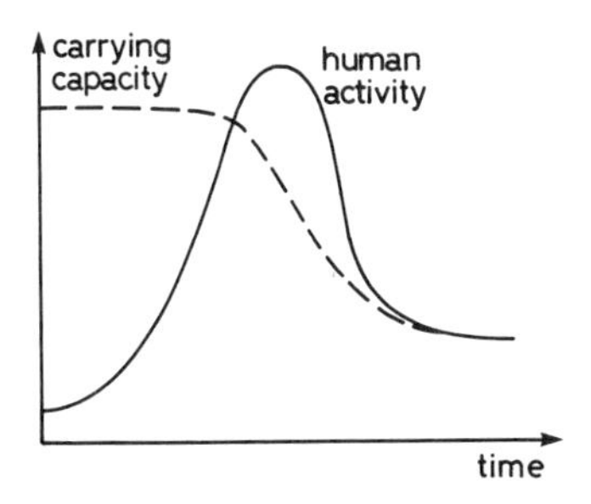

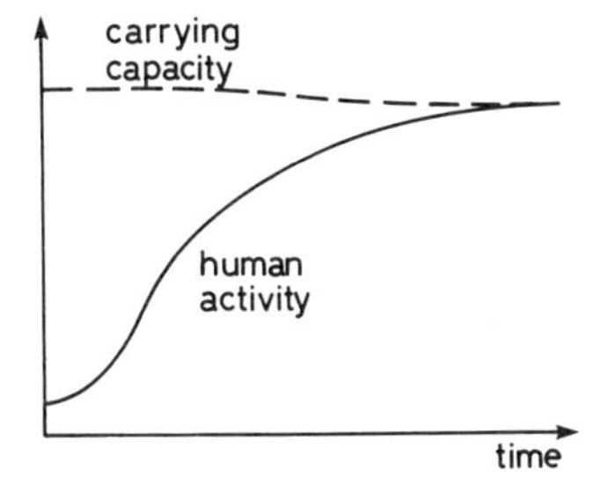

Figure 6.4
The reference mode for the study of material growth in a finite world

Figures 6.4a and b together represent the reference mode of the global modeling study.

After developing *organizing concepts*—human material activity, carrying capacity, and delayed response to the proximity of constraints—it became easier to discuss the question addressed. The concepts also facilitated verbal descriptions of the following group of processes judged responsible for the reference mode:

1. The level of human activity increases when there is room for expansion, that is, unutilized carrying capacity.
2. A sufficiently high level of human activity erodes the carrying capacity of the global environent.
3. There will be no response in the form of deliberate reduction of an excessive load until after a delay spent in data gathering and institutional change.
4. Exceeding the carrying capacity forces an involuntary downward pressure on human activity, for example, through starvation.

The system defined by these interactions can be represented by the causal diagram in figure 6.5. The causal diagram depicts the *basic mechanisms* of the World 3 model of growth toward finite limits, which are the smallest set of feedback processes considered sufficient to generate the reference mode in figure 6.4.

Taken together, the reference mode and the basic mechanisms constitute the "dynamic hypothesis" of the study. An early task in the modeling process was to test the dynamic hypothesis, that is, to check whether the basic mechanisms actually did generate the reference mode.

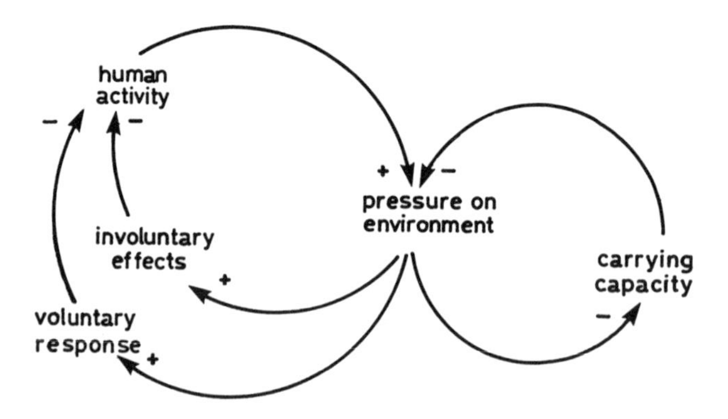

Figure 6.5
The basic mechanisms thought to underlie the reference mode in figure 6.4

An initial model was made for this purpose. The following levels (state variables) were considered sufficient to describe the system under study: human activity, carrying capacity, and voluntary response to environmental pressure. The simple model shown in figure 6.6 proved capable of reproducing the reference mode, as can be seen in figure 6.7. This initial model was the output of the conceptualization stage and the basis for developing an improved model.

6.5 Generalized Testing as a Guide in Model Improvement

In the same way that conceptualization is an iterative process, the subsequent formulation of the formal computer model proceeds in a recursive manner. Starting from a simple initial model, encompassing little more than an aggregate description of the basic mechanisms, the modeler gradually moves toward a more complete description. Thus the model becomes "better" through repetitive testing and correction of weaknesses.

But what criterion should be used to identify weaknesses? In the physical sciences, models are improved by comparing model-based point predictions with quantitive observations. This procedure is not optimal for improving relatively simple models of complex social systems. Simple models of complex systems exhibit a significant stochastic element, so there is little guidance to be obtained from comparison of detailed model prediction with specific real-world observations. In upgrading social system models, it proves better to employ a much broader set of model tests. In what could be called "generalized testing," all aspects of the model, not only model predictions, are tested, using all available knowledge as well as quantitative data.

Tests need not be restricted to only one characteristic of the model, such as its ability to predict events for which quantitative data are available. A large number of models are capable of reproducing any given time development. Therefore a descriptive dynamic model is not necessarily useful only because it reproduces observed historical behavior. Dynamic models have several other attributes that can be tested, including

- the capacity of the model to generate behavior modes corresponding to those observed or expected in the real world,
- the plausibility of the causal mechanisms chosen to represent the real world,

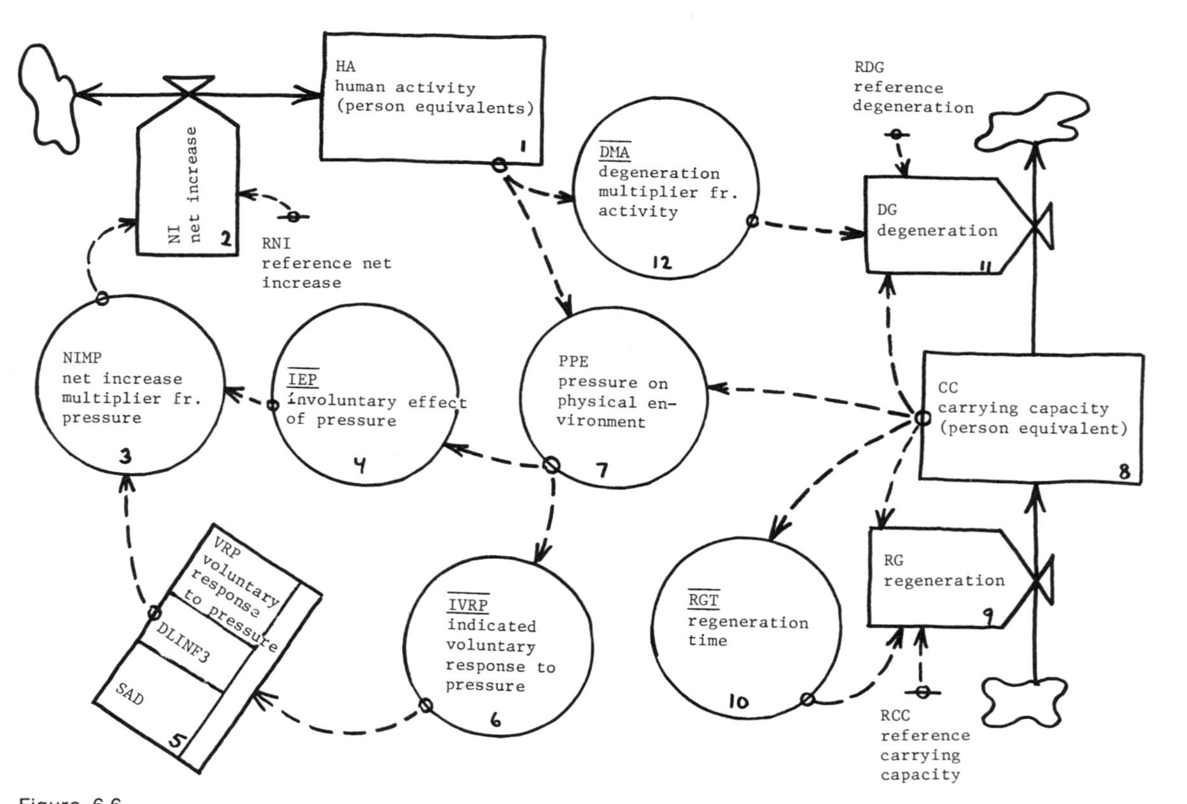

Figure 6.6
DYNAMO flow diagram for the initial model underlying the World 3 model

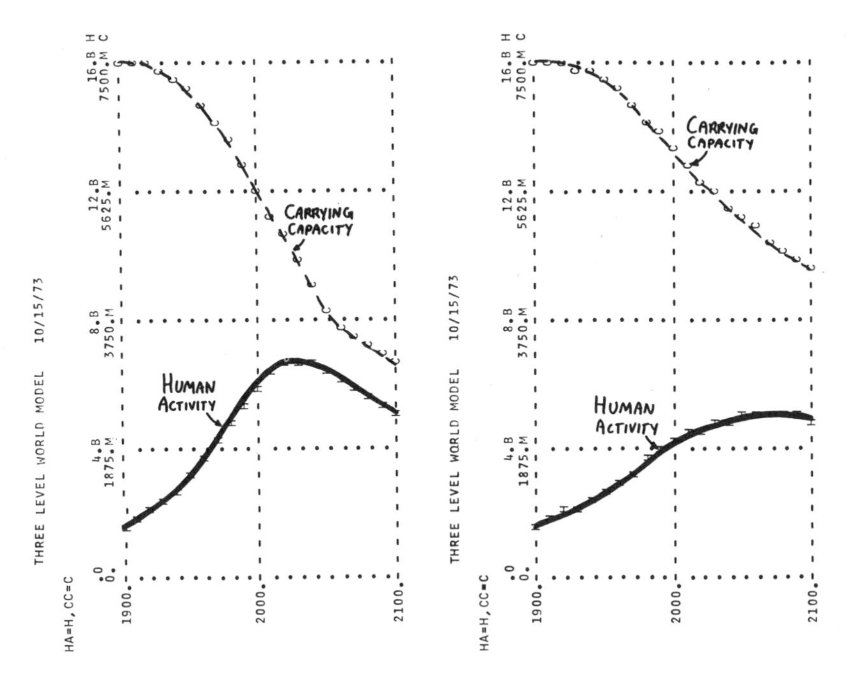

Figure 6.7
Runs of the initial world model: (a) overshoot and collapse and (b) anticipatory action

- the plausibility of the numerical values of model parameters,
- the compatibility of individual assumptions with established knowledge,
- the internal consistency of the full structure,
- the completeness with which the model includes the mechanisms thought to generate the problem addressed.

Models can be subjected to generalized testing at all stages of their construction, and revisions should be undertaken whenever the models fail to satisfy some criterion.

In judging how well a model meets the listed criteria, the modeler should not restrict himself to the small fraction of knowledge available in numerical form fit for statistical analysis. Most human knowledge takes a descriptive nonquantitative form, and is contained in the experience of those familiar with the system, in documentation of current conditions, in descriptions of historial performance, and in artifacts of the system. Model testing should draw upon all sources of available knowledge.

Generalized testing is a rigorous testing procedure. Still, a model structure that satisfies the tests is not an incontestable description of reality; nor is it the only model. On the other hand, it is certainly not a random accumulation of assumptions, since most conceivable structures would be eliminated by one or more of the criteria. Having survived generalized testing, a model acquires a certain stature and is ready for as many additional, preferably rigorous, tests as time and interest will sustain.

The result of practical application of generalized testing can be illustrated by the eventual fate of the initial world model (shown in figure 6.6). After successful tests of the dynamic hypothesis, the modelers began a long process of discovering and correcting errors and weaknesses, and extending and elaborating the original model to obtain an increasingly "better" model. Ultimately, they arrived at a model that appeared sufficiently credible to warrant experimentation to devise improved policies for managing the world system. Approximately twenty person-years of gradual extension and elaboration of the initial model led to the version of the model called World 3, shown in figure 6.8, that was published. Figure 6.9, showing two World 3 runs, shows how the reference mode is still intact even though World 3 has twenty-one levels versus three in the initial model. The basic structure of the expanded World 3 model is still similar to the initial model in figure 6.6, although the World 3 structure is less aggregate, more realistic, and more confusing.

6.6 The Recommended Procedure

The recommended procedure for model construction is summarized graphically in figure 6.10. No amount of prior lessons will transform modeling into a sequential execution of a set of activities requiring no repetition. The self-corrective mechanism of recurring testing and corrections of flaws represented by the narrow oscillating curve is in fact desirable during conceptualization and formulation, as long as the number of iterations remains reasonable. The recommended approach is merely designed to reduce the number of futile iterations by imposing some structure (represented by the broad band) on the process.

To summarize, the modeling process is split in a conceptualization and a formulation stage. The goal of the conceptualization stage is to arrive at a rough conceptual model capable of addressing a relevant problem. The formulation stage should embrace two processes: the test of the dynamic hypothesis, which is a preliminary check to see that the basic mechanisms included in the conceptual model actually reproduce the reference mode, and model improvement, which extends and elaborates upon the initial model until it is sufficiently versatile and detailed to serve the intended purpose.

The modeler should begin by selecting a process (observed or hypothetical, taking place through time) to represent the focus of the study. The chosen process should be described in terms of the time-varying behavior of certain key variables, and sketched on a graph. Only the most general features of the behavior should be recorded. The resulting reference mode serves as an approximate picture of the expected output of the initial model. The reference mode is not necessarily restricted to one time pattern; several characteristic behaviors may be required to define properly the problem focus. For models of past phenomena, the reference mode could consist of the historically observed behavior which the initial model should reproduce. The reference mode helps the modeler define the problem with greater clarity—it determines the time horizon of the study, and it indicates the necessary level of aggregation and the extent of the system boundary in the model.

Having specified the reference mode, the modeler should identify the fundamental real-world mechanisms assumed to produce the reference mode. He should select and describe the smallest set of feedback loops considered sufficient to generate the reference mode, that is, select the basic mechanisms. Forcing himself to express his assumptions in writing is an excellent way for the modeler to get rid of nonessential mechanisms and detail. A quick sketch of the basic mechanisms in causal diagram form may focus the modeler's thoughts and help him to visualize the system boundary. The sketch should be very simple (few loops), describing only fundamental mechanisms.

The dynamic behavior of interest—the reference mode—and the related causal structure—the basic mechanisms—determine in a precise way the aspect of reality to be studied. The reference mode helps the modeler focus on a specific phenomenon instead of ending in diffuse mapping of a system. Inclusion of the basic mechanisms forces the modeler to address a meaningful whole at all stages of model improvement. Subsequent models simply describe in more detail the fundamental relationships already present in the initial model.

The belief that the basic mechanisms can actually reproduce the reference mode remains an assumption until model simulation proves this dynamic hypothesis to be correct. The modeler should therefore build an initial model, consisting of the basic mechanisms, and simulate

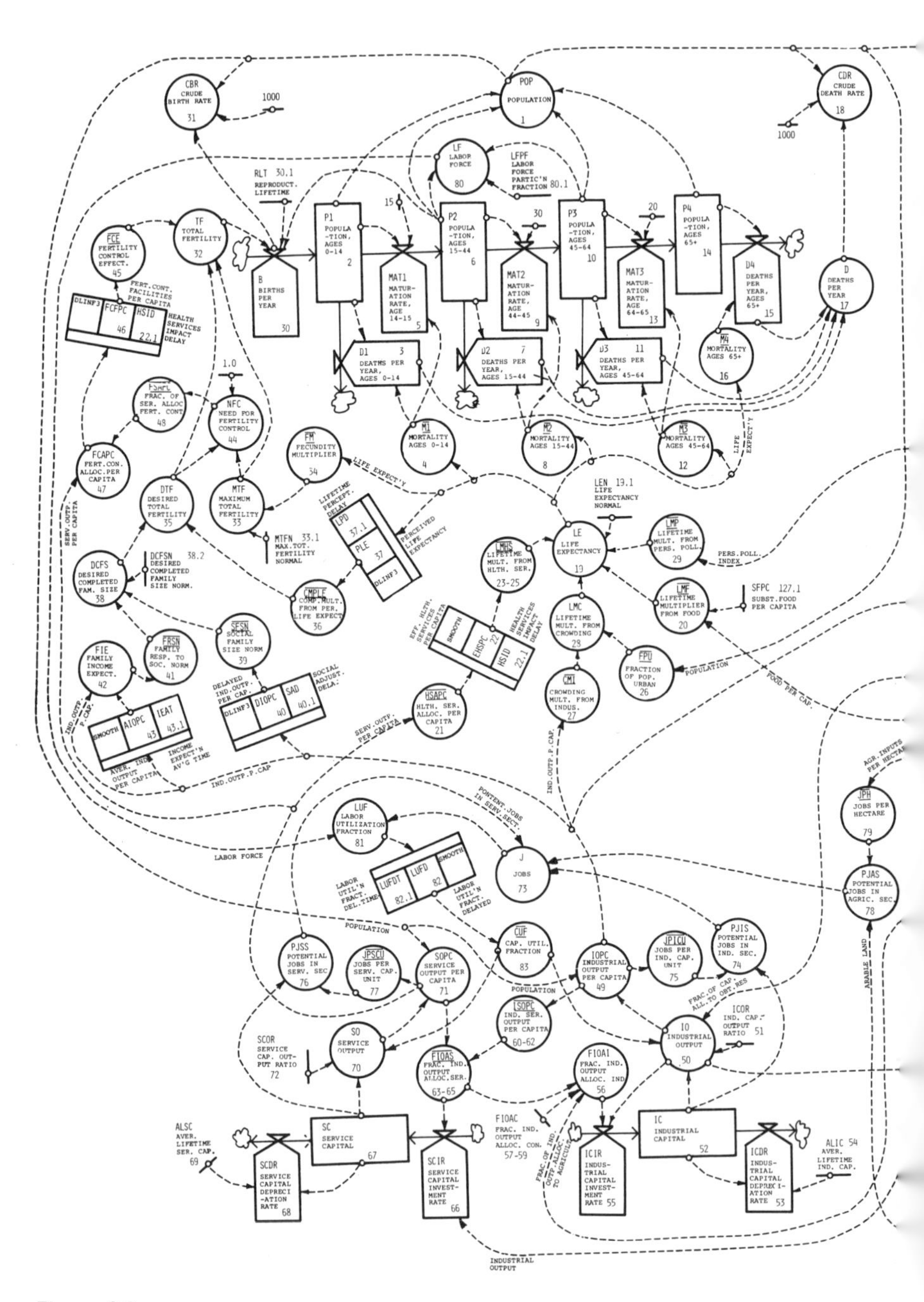

Figure 6.8
DYNAMO flow diagram for the World 3 model

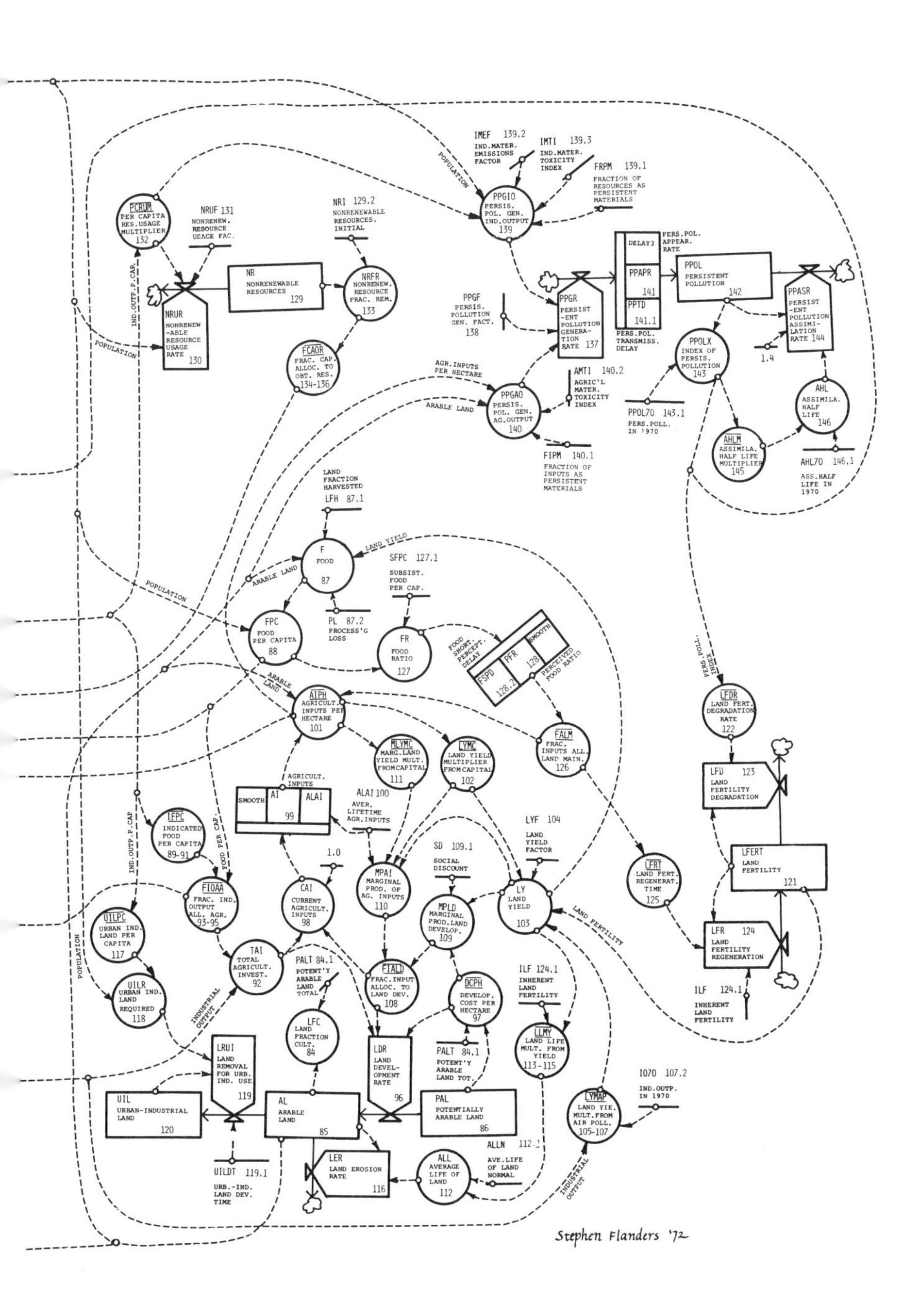

Stephen Flanders '72

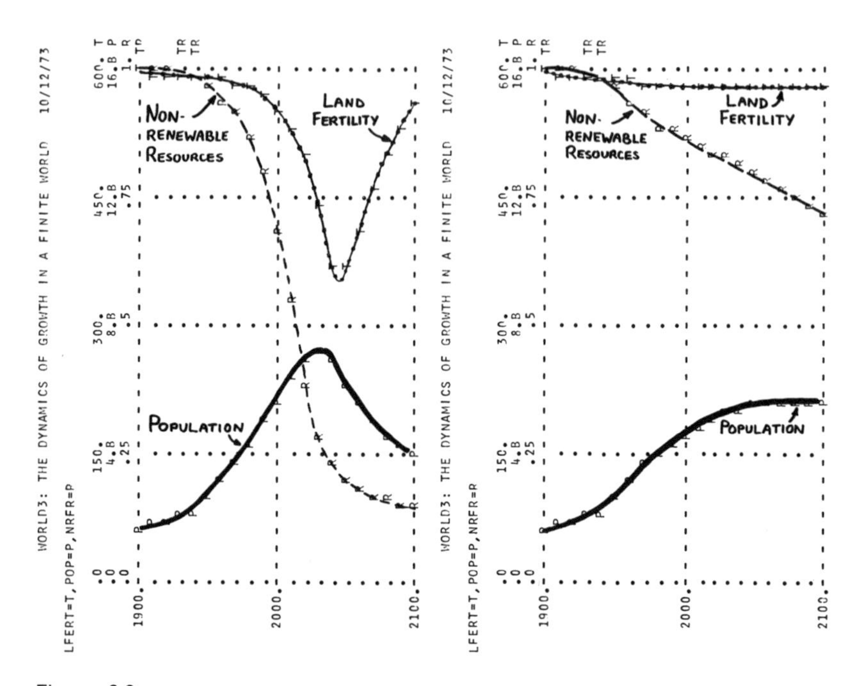

Figure 6.9
Runs of the World 3 model: (a) overshoot and collapse and (b) equilibrium through anticipatory action

(run) it to test the dynamic hypothesis—that is, to check whether the basic mechanisms can actually generate the reference mode.

The first step in formulating the initial model should be identification of the system levels. The levels describe a set of independent variables, together sufficient to describe the state of the system. The modeler begins by compiling a sufficient and possibly redundant list. To extract a set of levels, he continually eliminates variables that are not independent of elements already chosen as levels. After identifying the rates that govern change in the chosen levels, the modeler should add the causal influences on the rates. These causal influences should embrace the basic mechanisms that the model is supposed to include. The modeler should be able to construct a DYNAMO flow diagram at this point.

Next, the modeler should choose numerical values for table functions and time constants. Without belaboring the activity, he should then subject the completed structure to a first set of tests with respect to consistency, completeness, and reasonableness in its individual assumptions. If found satisfactory, the initial model should be run to determine

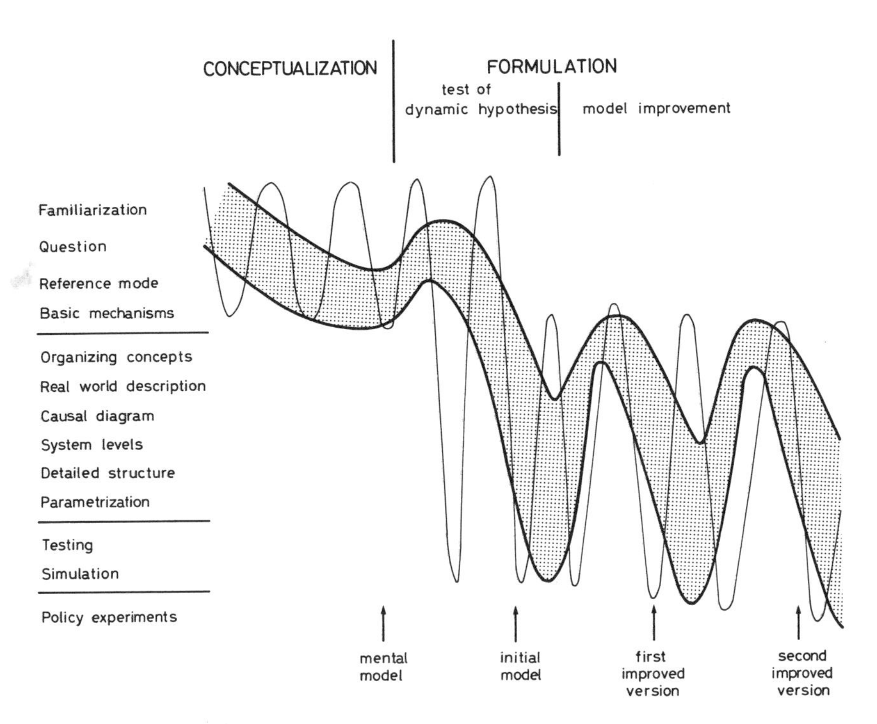

Figure 6.10
The recommended modeling procedure

whether it actually reproduces the major characteristics of the reference mode. If the model fails, flaws must be corrected in a new iteration. A new iteration may involve retracing all steps, even to the beginning with an altered problem definition. When the initial model passes generalized testing at an acceptable level, it is worthy of entry into the improvement stage.

The improvement stage consists of a never-ending series of extensions and elaborations to increase model richness or realism through changes in system boundary, level of aggregation, or detailed formulation. In most cases, improvement means making the model more complex. Since all models should be transparent, care must be taken to include new relationships only when they are necessary for adding a desired behavior mode, testing the effect of a policy, or attaining credibility with the user. The enrichment process must not be pursued to the point where the modeler can no longer grasp the connection between model assumptions and model output. During the improvement stage, the modeler may encounter powerful organizing concepts that make possible the reformulation of the whole study in a simpler, more elegant form. Such concepts, a valuable by-product of modeling, should be actively sought at all stages of the modeling process.

The reference mode acts as a catalyst in the transition from general speculation about a problem to an initial model that can later be left for routine improvement. This transition is the major creative step in modeling. Once the initial model is attained, the value of the reference mode in guiding progress diminishes. The models obtained later, by improvement, will show a richer variety of behaviors than the original reference mode.

Finally after extensive improvement, leading to a credible model structure and parametrization, the modeler may perform the policy experiments upon which his conclusion will rest. Conclusions should always be presented along with the model premises on which they are based. The premises may be organized in the more easily understood causal diagram format.

6.7 Common Mistakes and Some Guidelines

The modeling procedure outlined in this chapter may seem trivial to the novice. However, the value of an explicit guide to model construction may become more obvious if one considers the following list of mistakes that most modelers make:

- *Tendency to ramble due to lack of an explicit goal.* The first task of any modeling study is to define the goal of the effort. Without a clear objective it is not possible to decide what to include in a model, what aspect of reality to focus on, and when the result is "good enough."
- *Tendency to make excessively complex models to avoid inadvertent omission of important elements.* The simplest and safest response to uncertainty about whether a variable is important is to include it in the model. The modeler thereby evades challenging his own ignorance in an attempt to select only the few important factors. He also avoids the accusation of omission.
- *Tendency to exclude too much detail subsequent to failures with overly complex models.* In response to initial failures with excessive complexity, most modelers make simpler models. Lacking guidelines for what constitutes sufficient detail, he may well overreact and thereby slow down his progress toward the proper amount of detail.

- *Tendency to contract the scope of the model to permit a complete respectable analysis.* If the modeler focuses on a narrow slice of reality, he can include all elements that are commonly viewed as relevant without running into excessive complexity. No difficult choices among variables and relations need be faced, and the study attains an air of impregnable completeness and respectability.
- *Tendency to stick to earlier formulations to justify the effort put into their development.* It is psychologically difficult for the modeler to abandon a line of approach in which he has expended great efforts, particularly if the approach originally generated promising results. No doubt, a certain persistence is valuable, but there is a danger that commitment built up through the initial conceptualization struggle will keep the modeler from changing his approach when necessary.
- *Tendency to overemphasize causal diagraming, since causal diagrams constitute a tangible result without the finality of a completed model.* Until something appears on paper, the modeler may feel that a study has been unproductive. The modeler may have many reasons for wanting to postpone completion of a model. He may hesitate to go through time-consuming computer programing for a model that is still not fully satisfactory. A completed model tends to become "sacred" and unchangable. A completed model is also a conspicuous target for criticism. A causal diagram, on the other hand, represents a convenient compromise. The diagram is visible proof of effort; it can be produced without much toil; and it is still clearly unfinished and therefore not so susceptible to criticism.
- *Tendency to become stalemated in unending formulation problems, actually brought about by a lack of understanding of the real world.* Unless he has thorough understanding of the real-world process under study, the modeler will not be able to extract the few powerful assumptions constituting a useful model. However, when encountering modeling problems, the modeler is easily trapped into believing that the modeling tools are lacking the capability to represent his understanding of reality. Unending, futile attempts at formulating some part of a model are symptomatic of a lack of knowledge of the real world, and time should rather be spent on obtaining more knowledge. Knowledge constraints are more apparent when using powerful, versatile modeling techniques.

The following ten guidelines can help to counteract the common tendencies toward error:[6]

1. Explicit description of the dynamic behavior of interest—the reference mode—and assumptions about its cause—the assumed basic mechanisms—are necessary prerequisites for successful model building.
2. The modeler should consciously look for organizing concepts that are powerful descriptors of the basic mechanisms.
3. A dynamic hypothesis is obtained through combination of historical (or hypothetical) real-world behavior and simple structures with known behavior. Ideas for a productive perspective on reality can be obtained from familiar organizing concepts and existing models.
4. The system boundary must be wide enough to encompass the basic mechanisms, that is, a set of feedback loops capable of endogenously generating nontrivial dynamic behavior over the time period studied.
5. The purpose of the initial model is not to predict, but to test the dynamic hypothesis.
6. The initial model should only contain the basic mechanisms needed to generate the reference mode; additional complexity should then be gradually incorporated until a sufficiently realistic and versatile model is obtained.
7. The model should be kept transparent, even subsequent to the initial modeling stage. (a) A relationship should only be included in a model if necessary to generate a desired behavior mode, to test effects of a policy, or to achieve sufficient realism to gain credibility. (b) Each model link should represent a stable, meaningful real-world relationship in which the modeler has confidence.
8. Reduce the amount of detail (depth), rather than scope (breadth), if model complexity must be reduced.
9. Causal diagrams should be used only for exploration in the initial modeling stage and for communication of the "final" model; the main modeling should be performed by choosing and linking levels.

Notes

1. The author's Ph.D dissertation, "Conceptualizing Dynamic Models of Social Systems: Lessons from a Study of Social Change," Sloan School of Management, MIT, September 1973, reviews existing literature on conceptualization.

2. Notice that I do not say that the reference mode should be viewed as sacred or unalterable. It may well be that simulation runs with an early model do not recreate the reference mode, and that the modeler's response will be to change the reference mode and not the model.

3. John Høsteland, 1978 "Lagersvingninger i celluloseindustrien," GRS–136, Resource Policy Group, Oslo.

4. Of course, well-known system dynamics structures and literature was of great help in this work. See D. L. Meadows 1970, *The Dynamics of Commodity Production Cycles* (Cambridge, Mass. Wright-Allen Press).

5. Further description of the model and its conclusions can be found in D. H. Meadows, D. L. Meadows, J. Randers, and W. W. Behrens, 1972, *The Limits to Growth* (New York: Universe Books), and in D. L. Meadows, W.W. Behrens, D. H. Meadows, R. Naill, J. Randers, and E. K. O. Zahn, 1974, *The Dynamics of Growth in a Finite World* (Cambridge, Mass.: Wright-Allen Press).

6. The guidelines are discussed in more detail in my Ph.D dissertation, "Conceptualizing Dynamic Models of Social Systems: Lessons from a Study of Social Change."

Part IV
Formulation

Parameter Estimation in System Dynamics Modeling

7

Alan K. Graham

7.1 Introduction

System dynamics modeling offers an attractive tool for policy evaluation. Policy alternatives can be simulated by computer; thus the policies can be tested under completely controlled conditions. A special programming language, DYNAMO, is widely available and simplifies the mechanics of computer simulation. But simulations require a model, and model building is still an art in many respects. There is no step-by-step procedure that automatically produces a useful model. However, as more and more models are created, some steps in the modeling process have become clearer, and procedures can now be formalized to some extent. This paper lays out the main techniques of one important step in system dynamics modeling: the estimation of parameters. It is assumed that the reader is familiar with system dynamics and the DYNAMO language.

The appropriateness of the various parameter estimation techniques depends upon the entire context of the model-building process. For example, parameter estimation is closely dependent on equation formulation and model testing. Graham (1978) discusses these interdependencies in detail, so they will be touched on only briefly here. This chapter is a taxonomy of estimation techniques; the types of data, assumptions, and procedures that characterize each technique are specified. The purpose is to help the reader choose an appropriate technique and to avoid pitfalls.

Throughout the exposition of techniques, examples are drawn from a simple model of urban housing, which portrays the growth and maturity of an urban residential area. The appendix gives the complete equations for the model, which is based on a model described in Alfeld and Graham (1976, ch. 6).

7.2 Estimation Using Data below the Level of Aggregation of Model Variables

Most parameters in system dynamics studies are estimated on the basis of descriptive information obtained from participants in the system being modeled. That information is more detailed than data that corresponds directly to model variables. There are two kinds of model variables: level variables, which aggregate a collection of items like houses into a level of houses, and rate variables, which aggregate a stream of events like the construction of a house into a rate of housing construction. By contrast, data below the level of aggregation of model variables characterize the individual members of a level, or the individual events within a rate. For brevity, such data will sometimes be referred to as "unaggregate data."

For an example of setting parameters using unaggregate data, consider an equation representing housing demolition:

```
HD.K=H.K/HL                                                   1, R
HL=66                                                         1.1, C
    HD        - HOUSING DEMOLITION (HOUSES/YEAR)
    H         - HOUSES (HOUSES)
    HL        - HOUSING UNIT LIFETIME (YEARS)
```

The model assumes a constant average housing unit lifetime, HL, so that every year, 1/HL of the houses, H, are demolished. HL can be estimated in many ways from unaggregate data; in none is the equation used in computing the estimate. Equation (1) is used only to define the function of HL in the model. One time-consuming approach to estimating HL would be to survey a number of houses that have been demolished, take their ages at the time of demolition, and average those ages to determine an average housing unit lifetime, HL. The information used to set the parameter concerns individual houses and their demolition, which the model variables aggregate into a level of houses, H, and their outflow rate of housing demolition, HD. Thus the data come from below the level of aggregation of model variables.

As another means of obtaining information about the lifetime of houses, the modeler can examine the ages of existing houses, and observe the age at which very few houses remain standing. Or the modeler can consult someone who has observed construction and demolition of houses closely, and ask that person how long houses typically last (or survive neighborhood changes or eminent domain proceedings.) Indeed, asking experts questions of wider scope is the basis of the popular Delphi method (Turoff, 1970). The modeler can sometimes obtain descriptive histories of particular neighborhoods that chronicle the successive waves of demolition and construction, and thus gain some idea of how long the previous houses in that neighborhood lasted. The modeler can also call upon his own experiences in observing the physical decay of one or more houses, and extrapolate to estimate how long a house will last. Finally, lacking any better information, the modeler can take two extreme values that are clearly too large and too small and pick a value somewhere between them. Modelers seldom rely on this last technique, since it is based on the implicit assumption that in the entire world there is no available information that would lead the modeler to a single estimate of the parameter in question.

These examples suggest that the sources for data below the level of aggregation of model variables are numerous and diverse. The data for the lifetime of houses came from city hall records, a history book, expert testimony, and the modeler's own day-to-day experiences. In fact, *all* factual knowledge about a system—records, books, eyewitnesses, and personal experience—falls into the category of unaggregate data; the only exception is collected statistics corresponding to model variables. Thus unaggregate data are by far the most abundant source of knowledge about real systems.

Table Functions

We have shown six different ways to estimate the average housing unit lifetime using unaggregate data. All six are straightforward. This section describes a technique for estimating the 5 to 15 numbers that typically specify a table function. This may seem to be a formidable estimation problem, but it can be broken into subproblems: estimating the value and the slope of the function at one extreme, at the normal value, and at the other extreme, and connecting those known values and slopes with a smooth curve. Once these four subproblems are solved, the table function is known to within a narrow range of values.

For example, consider a group of equations that represent the effect of land availability on housing construction. The rate of housing construction, HC, is proportional to the number of houses, H, already within the urban area; thus when other things remain the same, more houses, more infrastructure, and more people create a larger market for new housing construction. The parameter that gives the constant of proportionality is the housing construction normal, HCN. The rate of housing construction, HC, is modulated by land availability; this modulation is accomplished by the housing-land multiplier, HLM:

```
HC.K=H.K*HCN*HLM.K                                          2, R
HCN=0.07                                                    2.1, C
     HC      - HOUSING CONSTRUCTION (HOUSES/YEAR)
     H       - HOUSES (HOUSES)
     HCN     - HOUSING CONSTRUCTION NORMAL (FRACTION/YEAR)
     HLM     - HOUSING LAND MULTIPLIER (DIMENSIONLESS)
```

The housing-land multiplier, HLM, responds to land availability, which is quantified by the land fraction occupied, LFO:

```
HLM.K=TABLE(HLMT,LFO.K,0,1,.1)                              3, A
HLMT=.8/.95/1.075/1.2/1.3/1.35/1.35/1.25/1/.6/0             3.1, T
     HLM     - HOUSING LAND MULTIPLIER (DIMENSIONLESS)
     TABLE   - TABLE INTERPOLATION FUNCTION
     HLMT    - HOUSING LAND MULTIPLIER TABLE
     LFO     - LAND FRACTION OCCUPIED (DIMENSIONLESS)
```

The graph for HLM is shown in figure 7.1. To estimate the table function for HLM, first consider the extreme condition of zero land occupancy, where incentives for construction should be less intense than with higher occupancy. When the land fraction occupied, LFO, approaches 0 (near the left side of the curve in figure 7.1), most of the area being modeled is vacant land. The area's viability as a future city has not yet been demonstrated. Developers cannot rely on continuing demand for the housing units they construct. Also, services taken for granted in more heavily settled areas must be installed in each successive new neighbor-

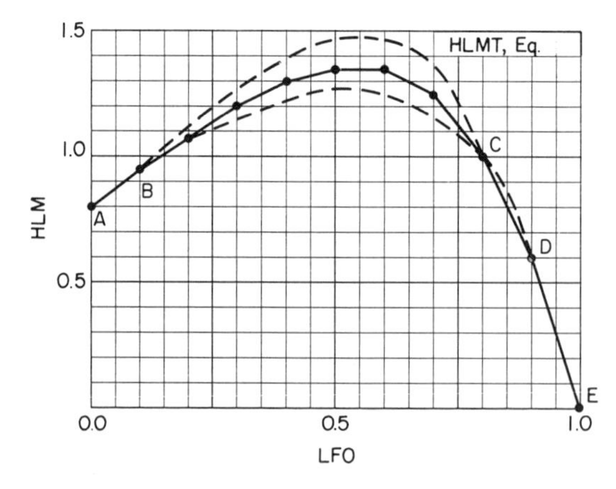

Figure 7.1
Housing-land multiplier table, equation (4)

hood: roads, sewers, electricity, gas, schools, and public transportation. Sparsely settled areas often cannot make city water or sewers (let alone public transportation) an economical proposition. So when the land fraction occupied LFO equals 0 the housing-land multiplier, HLM, should be lower than at most other values of LFO (point *A* on figure 7.1 uses the value of 0.8 for HLM).

Adding housing units to a sparsely settled area encourages more urban services. Urban services demonstrate the area's viability, and make housing construction more profitable. But adding a few houses cannot pay for the infrastructure—schools, roads, libraries, and utilities—necessary to deliver a complete ensemble of urban services. The curve for the housing-land multiplier table, HLMT, should slope upward, but not very steeply from where LFO equals 0 (see the line segment between points *A* and *B* in figure 7.1).

Now consider the normal condition. Equation (2) defines the rate of housing construction, HC, so that it occurs in normal proportion to houses (specified by housing construction normal, HCN) when the housing land multiplier, HLM, equals 1.0. For consistency, HLM must equal 1.0 at whatever value of the land fraction occupied LFO is defined as normal. Normal conditions are defined to occur near the end of the area's growth when land availability begins to constrain further construction. In this model, the normal value of LFO is defined to be 80 percent land occupancy (point C in figure 7.1).

At the normal condition, land availability begins to constrain housing construction. Thus the table function must have a negative slope at the normal point, so that diminishing land availability in the model likewise begins to constrain further construction in the model. The negative slope at the normal point in turn implies that the table function must exceed a value of 1.0 just under the normal value of land fraction occupied, LFO.

Now consider the other extreme condition in which the land fraction occupied, LFO, equals 1.0. The land area within the city or district being modeled is totally occupied; even the least desirable sites have been built upon. Regardless of the incentives to construct, no housing can be constructed within the area being modeled until there is some physical space available upon which to build—that is, until LFO is less than 1.0. So the housing-land multiplier, HLM, should equal 0 when LFO equals 1.0, which establishes point *E* in figure 7.1.

If the land fraction occupied, LFO, was not 1.0 but close to 1.0 (nearly full land occupancy), urban services like sidewalks, schools, libraries, roads, and public transportation would be fully developed. To be sure, the crowding and lack of desirable construction sites implied by an LFO close to 1.0 would not cause housing construction to take place as rapidly as under normal conditions. Nonetheless, any small reduction of LFO from 1.0 opens up the possibility of appreciable housing construction. Therefore the curve of the housing-land multiplier should have a steep slope as LFO approaches 1.0 (see the line segment between points *D* and *E* on figure 7.1).

So far, values have been estimated at and near two extreme conditions and at the normal condition. Now all that remains is to draw a curve through the estimated points. Any sharply bent or kinked curve is probably not realistic. A bend or kink implies something special about the exact conditions at which the bend or kink occurs. Since the housing-land multiplier table, HLMT, represents many phenomena (prices, availability, infrastructure, and so forth), the probability is small that all phenomena would show major changes at a unique set of conditions. Accordingly, the curve for HLMT (and in general for all highly aggregated relationships) should change smoothly.

Solving the subproblems of extreme and normal conditions, and connecting known points with smooth curves, allows the modeler to estimate a nonlinear table function with confidence. The dotted lines show alternate curves for HLMT that also satisfy the constraints for extreme values, slopes, and smooth curvature; these constraints allow little latitude in specifying HLMT. The estimated table thus summarizes observations of a large number of processes below the level of aggregation of model structure. HLMT is the aggregate representation of these processes and their effect on housing construction.

Ad Hoc Computation

It sometimes happens that no unaggregated data are available that correspond directly to the parameter being estimated. In the housing example, this would happen if no information were available about the ages of houses at demolition. However, other unaggregated data may exist that describe the process of demolition, and these data may suffice to infer an estimate of the housing unit lifetime, HL. This is an ad hoc computation used to estimate a parameter from unaggregate data.

For example, suppose the modeler walks through a district and observes that of a fairly homogeneous mix of housing units of different ages and types about one in a thousand is in the process of being demolished. Suppose also that conversation with a wrecking crew reveals that it takes about three weeks to demolish a single building when neighboring buildings are left intact. From this information, the fraction of the housing stock demolished each year can be computed: it takes three weeks to demolish a building, or $(3/52) = .0577$ years. Thus, if .001 of the housing stock is observed to be in the process of demolition, and if that rate of demolition continues for a whole year, the fraction of the housing stock demolished each year is $(.001/.0577) = .01733$. This fractional demolition rate (whose unit of measure is per year) is the reciprocal of the housing unit lifetime, HL (whose unit of measure is years). Thus HL is $(1/.01733)$ or 57.7 years, which is not an unreasonable estimate.

This example suggests that parameters can be estimated by ad hoc computations based upon readily available unaggregate data. These computations can take many forms; see Senge (1975), for example.

Pitfalls

The greatest single pitfall in using unaggregated data lies in formulating a model structure and parameters that are aggregated to the point where the processes characterized by the parameter values cannot be reliably observed. As a result the parameters have little real-life meaning, and to estimate them, the estimator must draw conclusions based on a mental model of the behavior of the system, rather than simply reporting observations. For example, in the housing model a variety of processes determine how long it takes the system to make a transition from growth to full land occupancy—incentives to construct housing, supply and demand effects in the land market, and housing depreciation, for instance. In the model, several parameters characterize such diverse processes. An alternative formulation of the model might have contained a single parameter that specified the time constant for the transition from growth to equilibrium. Urban experts may well be willing to estimate such a quantity, but the estimate would be a conclusion or opinion drawn from their mental models of how the cities behave, rather than a report on a single cause and effect relation within the city.

Another instance of confusing conclusions with observations might occur if the simple housing model described here is expanded to include demand for housing arising from the size of a population, but expanded improperly. A properly disaggregated way of modeling demand is to represent explicitly the growth of population and its response to housing availability. But the modeler may be tempted to simplify and aggregate the model structure, perhaps by attempting to represent the influence of population with a relationship that stimulates housing construction, HC, when the stock of housing units, H, has grown slowly (representing growth of population faster than housing and thus an increase in demand). Similarly, such a relationship might retard housing construction, HC, when housing units, H, has grown rapidily (representing overexpansion of the housing stock relative to the population and thus a slackening of demand). During times of moderate growth, this relationship would have a neutral effect on HC, representing the assumption that people could be found to occupy the houses; implicitly, the population growth would keep pace with the growth in housing.

What are plausible values for the parameters in this formulation? What is "moderate growth" relative to the speed of population movements? What should the magnitude of the effect of rapid growth on HC be? These questions cannot be answered by first-hand observations of cause and effect relationships; the questions call for conclusions based on mental models of the dynamics of the city. Can the modeler predict from intuition alone and characterize with one delay and one table function the dynamic interactions among housing and population, incorporating births, deaths, incentives for migration, family formation, or the ability of construction companies to expand? If not, the model is too aggregated for parameters to be estimated reliably from the unaggregate data.

Two possible actions can be taken when the parameters of a model are too aggregated to be set reliably from available unaggregate data. One course is to use another estimation technique (usually a statistical technique) and data at the level of aggregation of model structure. It seems unwise, however, to attempt to estimate a simple relationship if the actual system is so complex that expert opinion is unreliable. A preferable course of action is to restructure (usually disaggregate) the model so that its parameters correspond directly to observable, unchanging characteristics of the system. The disaggregation usually involves not only subdivision of levels into more levels, but also explicit addition of feedback loops that control the levels. For example, consider the relationship between land availability and urban housing construction discussed at the beginning of this section. Mass (1974) and Miller (1975) disaggregate this relationship to portray the details of land pricing, speculation, rezoning, and land use.

In summary, unaggregate data are by far the most abundant source of information. There is a wide range of specific estimation techniques, extending from direct observation to ad hoc computations based on direct observation. Of particular interest is the problem of estimating the parameters of a table function. This problem can be reduced to the subproblems of estimating extreme values and slopes, specifying the normal point, and drawing a smooth curve through the extreme and normal points. The main pitfall in estimation with unaggregate data is formulating an equation and its parameters in an aggregate, simplified manner, so that participants in the system cannot reliably observe a value of the parameter as a characteristic of the real system.

7.3 Estimation Using a Model Equation

A model equation and its parameters specify a relationship between two or more variables. Estimation using a model equation starts with statistics that aggregate individual items or events that correspond to model variables. From such data the modeler derives the parameter values that enable the model equation to match the "real" relation between the variables. Estimation using a model equation encompasses all single-equation regression techniques. Theil (1971) offers a general treatment of regression techniques. Hamilton (chapter 8) and Mass and Senge (chapter 10) discuss the application of these techniques to system dynamics models.

For an example of estimation using a model equation, consider that the housing model calculates the rate of housing demolition, HD, as the number of houses, H, divided by the average housing unit lifetime, HL. Thus, if data are available for HD and H, HL can be estimated from the model equation:

$$HD.K = H.K/HL$$

$$HL = H/HD.$$

Estimation using a model equation is less frequent in system dynamics studies than estimation from unaggregate data. Nevertheless, two forms of model-equation estimation—one involving conversion factors, the other fractional rates of flow—have been useful in many studies and are therefore discussed here.

Conversion Factors

Many model parameters are conversion factors: they convert quantities from one dimension to another. For example, land per house, LPH, converts housing units to an equivalent number of acres. Equation (4) uses LPH in the definition of land fraction occupied, LFO:

```
LFO.K=(H.K*LPH)/AREA                                        4, A
LPH=0.1                                                     4.1, C
AREA=9000                                                   4.2, C
    LFO     - LAND FRACTION OCCUPIED (DIMENSIONLESS)
    H       - HOUSES (HOUSES)
    LPH     - LAND PER HOUSING UNIT (ACRES/HOUSE)
    AREA    - LAND AREA (ACRES)
```

These equations can be manipulated to compute the parameter as a function of the real data, in this case, the real LFO, H, and AREA:

$$\mathrm{LPH} = \mathrm{LFO} * \mathrm{AREA}/\mathrm{H}.$$

Estimating conversion factors offers a straightforward means of ensuring that the absolute magnitudes of model variables are realistic. Schroeder and Strongman (1974) show how real data were used to estimate conversion factors for a model of Lowell, Massachusetts. That model exhibits realistic magnitudes for population, housing, and employment.

Normal Fractional Rates of Flow

Equation (2) defines the rate of housing construction, HC, in terms of a level (houses, H), a normal fractional rate of flow (the housing construction normal, HCN), and a dimensionless multiplier (the housing-land multiplier, HLM). This format

$$\text{rate} = \text{level} * \text{normal fraction} * \text{multipliers}$$

is widely used (Alfeld and Graham, 1976, pp. 123–126, provides further discussion). One reason is that the multiplier can easily be estimated when it is normalized around 1.0 (see section 7.2). This format also facilitates the estimation of the normal fraction:

$$\text{normal fraction} = \text{rate}/(\text{level} * \text{multipliers})$$

Under normal conditions (however defined), the multipliers, by definition, assume values of 1.0. So the normal fractional flow rate can be computed by dividing the observed rate by the observed level, both measured during a period of normal conditions. For example, suppose that the year 1960 is defined as the normal period for the urban area being modeled. Then, if the data are available, a value for housing construction normal, HCN, is obtained from the number of housing units constructed in the area during 1960 divided by the number of housing units in the area in 1960.

Pitfalls

Data at the level of aggregation of model variables must be collected for specific purposes. This fact may be a pitfall when data are collected for purposes different from those of the model. The completeness of a set of variables, the definition of the aggregated variables, or the time frame may render data unsuitable for estimating parameters using a model equation.

For example, the definition of the collected data may be inconsistent with model definitions. The housing-land multiplier, HLM, is defined so that housing construction is impossible when land fraction occupied, LFO, equals 1.0. Thus LFO must reach 1.0 when the area has all the housing it can hold. This means that land per house, LPH, must include not only the land directly beneath each housing unit but also the adjacent land for yards, sidewalks, driveways, roads, schools, and stores. The land per house, LPH, for a particular area could be calculated from the land area zoned for residential use (minus the area of vacant lots) divided by the number of dwelling units within the area. However, land in many cities is zoned for both residential and commercial use; some fraction of that land must be included in the residential land area as well. So the modeler must be aware of exactly what is and is not counted to form a particular piece of data, and whether that definition is consistent with the way corresponding quantities are used in the model.

Another difficulty with inappropriate data occured in an attempted revision of Forrester's (1969) urban model by Babcock (1970). Babcock set the normal constants with data on levels and rates of flow, but not with data only for the normal period; data for cities near equilibrium were used also. The simple housing model presented here can show what happened as a result. The housing model reaches equilibrium after a period of growth in the housing stock: the housing stock grows until a shortage of land suppresses further housing construction. Because the normal conditions in the model are growth conditions, the housing-land multiplier, HLM, must suppress housing construction by going well below 1.0. Suppose the actual rate of housing construction, HC_a, is divided by the actual number of houses, H_a, to obtain a computed value for the housing construction normal, HCN_c. Assuming the model equations are accurate, but using actual equilibrium data to compute HCN:

$$HCN_c = \frac{HC_a}{H_a} = \frac{H_a * HCN_a * HLM_a}{H_a} = HCN_a * HLM_a.$$

Since HLM < 1 in equilibrium, then

$HCN_c < HCN_a$.

The computed value of HCN, if used instead of the actual value, reduces the model's impetus to grow, and thus reduces the extent to which HLM must drop to bring the model into equilibrium. Similarly Forrester (1969) holds that growth ceases when land shortage and unfavorable internal conditions (principally a job shortage and predominance of lower-income groups) depress construction. Using data from near-equilibrium to compute normal fractions considerably reduces the extent to which internal conditions in the model must decline to halt growth. In fact, the model will no longer reproduce and account for depressed urban conditions. Babcock's re-estimated model therefore no longer fulfills its purpose, merely because the implicit assumptions used in parameter setting are violated.

The pitfall then in estimating parameters with data at the level of aggregation of model variables is that the computations require two assumptions: accuracy of an equation and appropriateness of the data. Such assumptions always constitute "more rope to hang yourself with."

7.4 Estimation Using Multiple Equations

As just described, estimation using a model equation consists of manipulating the equation to compute a parameter value. By contrast, estimation using multiple equations consists of manipulating several equations to compute a parameter value. Both techniques use data at the level of aggregation of model variables. The two techniques are distinguished here because they are usually distinct in practice. Usually either one equation is used analytically or all equations are used in simulations. Also, the pitfalls tend to be rather different.

For example, the housing construction normal, HCN, can be estimated by finding the value of HCN that causes housing growth to fit the observed rate of growth. This estimation would use all the equations. The fitting could be performed either with repeated simulations or, if possible, by a computation. For an example of such a computation, suppose that the stock of housing grows at 5.5 percent per year under normal conditions. Also suppose that, from observation of housing demolition, the housing unit lifetime, HL, is estimated to be 66 years—that is, 1/66 of the houses are demolished each year. If the model equations and the parameter HL are assumed to be correct, then the housing construction normal, HCN, must exceed 1/66 by 0.055 to produce the observed rate of growth during the normal period. Therefore HCN can be inferred to be $1/66 + 0.055 \cong 0.07$. Although this computation uses all of the model equations, it can be performed simply because the model is very simple; more complex models usually require more elaborate numerical computations.

For another example, suppose a real system exhibits fluctuations of some specific period. The modeler can choose the magnitudes of time constants of the system so as to produce oscillations near the real period. Forrester (1968, ch. 10) derives a simple rule of thumb: for a second-order undamped system with two time constants T_1 and T_2, their geometric average approximately equals the period divided by 2π: $\sqrt{T_1 T_2} = P/2\pi$.

Just as the estimation using one model equation subsumes single-equation regression, so estimation using multiple equations subsumes a family of statistical techniques. The most general technique is full-information maximum likelihood (FIML) estimation. Unfortunately, for nontrivial problems FIML usually requires extravagant computation time. Therefore two families of less general techniques have evolved. One family usually requires linear formulations, and information on exogenous variables that is both complete and accurate. These are the multiple-equation regressions (Theil, 1971, ch. 9–10). They are much more efficient computationally than FIML. The other family of techniques restrict the models to be dynamic (nonsimultaneous) and only mildly nonlinear. These are the full-information maximum likelihood via optimal filtering (FIMLOF) techniques derived from control theory (Schweppe, 1973). Although they require significant computation, they offer conceptual simplicity as well as estimating nonlinear dynamic models with flawed information on only a subset of model variables (Peterson, 1975; see also chapter 11 in this volume). Software is available commercially for doing FIMLOF estimations with system dynamics models (Peterson and Schweppe, 1974).

The general pitfall of multiple-equation estimations is the same as for single-equation estimations: the techniques assume the accuracy of the equation(s) and the data. The implicit assumption that most often thwarts multiple-equation estimation is that the discrepancy between real behavior and model behavior is due to the values of the parameters being estimated. In other words, the discrepancy can be misattributed. In the oscillation example cited above, if T_1 is inaccurate, and T_2 is being estimated , the estimate of T_2 will also be inaccurate as well in order that $\sqrt{T_1 T_2} = P/2\pi$. A subtle type of misattribution sometimes occurs during model testing: if the model exhibits unrealistic behavior, the modeler changes a parameter to eliminate the unrealistic behavior. Thus the parameter has been estimated in the sense that its value has been chosen to allow the model equations to generate behavior that matches observed behavior. The problem is that the unrealistic behavior of the model may have been due not to the inappropriate parameter value but to an unrealistic formulation or to some *other* parameter value being awry. To attribute the unrealistic behavior to the original value of the altered parameter may indeed produce the right behavior but for the wrong reasons.

Another pitfall of both single-equation and multiple-equation parameter estimation arises from their use of data at the level of aggregation of model variables. To varying extents, both techniques force magnitudes of model variables and the relationships among them to conform to the magnitudes and relationships in the data. The forced conformity of some aspect of behavior to real data preempts the comparison of behavior to data as a validity test. For example, if the modeler estimates model parameters from data (at the level of aggregation of model variables) for housing units, housing construction, housing demolition, and land occupancy, the model is likely to replicate the overall behavior of the housing stock. But confidence in the model would be greater if the parameter estimation used unaggregate data and still resulted in a model that replicates aggregate behavior. In other words, if the model replicates real behavior when it doesn't have to, the replication is another basis for confidence in the model. Estimations from data at the level of aggregation of model variables hinders the modeler in using such a validity test.

One way to circumvent this pitfall, if there are enough aggregate data, is to follow the common econometric practice of using only part of the data from estimation and the rest for validation (Theil, 1971, pp. 603–604). This strategy is sometimes unworkable when not enough data exist, or if model equations are not general enough to replicate more than one set of data. The other way to avoid the pitfall is to use unaggregate data to estimate parameters and reserve aggregate data for validation; the latter strategy is commonly followed in system dynamics studies.

7.5 Conclusions

Three general categories of parameter estimation techniques have been presented. Estimation using data below the level of aggregation of model variables relies on observations of individual items or events that are represented in the aggregate by model variables. The principle pitfall of this technique is structuring the model on a level of aggregation too high to allow observers within the system to reliably translate their experiences into parameter values.

The two other techniques are estimation using a single model equation and estimation using several or all model equations. Both techniques assume the correctness of the given equation(s); they use the equation(s) to infer parameter values from data corresponding to model variables. These techniques share pitfalls. First, use of data at the level of aggregation of model variables diminishes the ability to validate. Second, these techniques are vulnerable to systematic errors when assumptions are violated. Econometricians encounter this pitfall in estimating simultaneous-equation models. Even though multiple-equation methods theoretically deliver greater accuracy than multiple applications of a single-equation method, the multiple-equation methods are more sensitive to minor violations of assumptions (less robust) than single-equation methods (Theil, 1971, p. 552). Similarly, parameter estimation from data at the level of aggregation of model variables is less robust than parameter estimation from data below the level of aggregation.

How should the modeler choose among the three techniques? As a point of departure, the modeler need not favor the equation-based techniques over the use of unaggregate data on the basis of accuracy: Senge (1978) shows that estimation from unaggregate data can be as accurate as other techniques. Moreover, the pitfalls or limitations of the three estimation techniques are only one consideration in choosing among them. The appropriateness of a given technique also depends on the context of the model-building effort, most notably on how the model variables have been selected, and how the model will be tested.

Variables may be selected by several criteria. Variables selected on the basis of their ability to contribute to point-predictive accuracy suggest the use of aggregate data and highly aggregated relationships. By contrast, variables in system dynamics studies are usually chosen because they can reproduce the causes of the problem being analyzed, and because they can be recognized and validated by participants in the system being modeled. These considerations favor the use of variables at a level of aggregation close to that observable by individuals, without regard to whether or not statistics are available. Goals and psychological pressures are seldom measured but can be central in the model's description of why problems arise. Thus a typical system dynamics model is formulated in a way that makes estimation from unaggregate data reliable and that circumvents the pitfall of highly aggregated, unobservable relationships.

What tests must the model pass for it to fulfill its purpose? If the tests concentrate on the ability of the model to predict future values, then the parameter estimation techniques used must emphasize making the model variables fit aggregate data. However, the situation is different in most system dynamics studies. The ultimate aim is to predict the qualitative results of a policy change. Such predictions are difficult or impossible to evaluate quantitatively and directly. Instead, system dynamics studies tend to use a broad array of tests of structure, behavior, and policy impact (Forrester and Senge, 1978). Many of these tests do not depend on having data at the level of aggregation of model variables; hence the modeler can be flexible in choosing a parameter estimation technique. Specifically, by setting parameters from unaggregate data, the modeler can reserve any aggregate data for the purpose of validity testing—comparing model behavior to real behavior. This reservation of data is analogous to the common econometric practice of estimating parameters from data from one time interval, and testing the results with data from another time interval.

A final aspect of the relation between parameter estimation and model testing concerns sensitivity assessment: much of the simulation in a typical system dynamics study aims at identifying the equations and the parameters that are central in producing the behavior and policy results. In a model of realistic detail, only a few parameters can alter the outcome of the model if they are changed. (Graham, 1978, details the process of identifying and dealing with sensitive parameters.) Thus it is common practice to set parameters on the basis of information at hand (usually unaggregate data) and defer intensive data collection and parameter estimation until model testing reveals the parameters that require such measures.

What techniques should the modeler use to estimate the parameters of a model? The estimation techniques should facilitate, and be facilitated by, the other phases of the modeling effort: the model should be formulated in such a way as to avoid the pitfalls or limitations of the estimation techniques. The model testing should guide estimation efforts.

And information should be used in estimating in a way that facilitates validation. These considerations are reflected in the following ensemble of recommended system dynamics practices:

1. Use a model structure that is detailed and realistic enough to allow participants in the system to supply data below the level of aggregation of model variables.
2. Whenever possible, estimate parameters with data below the level of aggregation of model variables and reserve data at the level of model variables for validity testing.
3. Use techniques based on model equations only as secondary techniques since they are vulnerable to systematic error.
4. Use simulation to identify the equations and parameter values that are critical to the outcome of the modeling effort and focus subsequent efforts on those equations and parameters.

References

Alfeld, Louis Edward, and Alan K. Graham. 1976. *Introduction to Urban Dynamics.* Cambridge, Mass.: MIT Press.

Babcock, Daniel L. 1970. "Analysis and Improvement of a Dynamic Urban Model." UCLA Ph.D. dissertation. Available from University Microfilms, Ann Arbor, Mich.

Forrester, Jay W. 1969. *Urban Dynamics.* Cambridge, Mass.: MIT Press.

Forrester, Jay W., and Peter M. Senge. 1978. "Tests for Building Confidence in System Dynamics Models." System dynamics group working paper D-2926-5, MIT, Cambridge, Mass.

Goodman, Michael R. 1974. *Study Notes in System Dynamics.* Cambridge, Mass.: MIT Press.

Graham, Alan K. 1978. "Parameter Formulation, Estimation, and Assessment in System Dynamics Modeling." System dynamics group working paper D-2349-1, MIT, Cambridge, Mass.

Mass, Nathaniel J. 1974. "A Dynamic Model of Land Pricing and Urban Land Allocation." In *Readings in Urban Dynamics,* vol. 1. Edited by Nathaniel J. Mass. Cambridge, Mass.: MIT Press, 1974.

Miller, John S. 1975. "Urban Dynamics and Land Rezoning." In *Readings in Urban Dynamics,* vol. 2. Edited by Walter W. Schroeder, et al. Cambridge, Mass.: MIT Press.

Peterson, David W. 1975. "Hypothesis, Estimation, and Validation of Dynamic Social Models." Ph.D. dissertation, MIT, Cambridge, Mass.

Peterson, David W., and Fred C. Schweppe. 1974. "Code for a General Purpose System Identifier and Evaluator (GPSIE)." *IEEE Tr. Auto. Control AC–19:* 852.

Schroeder, Walter W. III, and John E. Strongman. 1974. "Adapting *Urban Dynamics* to Lowell." In *Readings in Urban Dynamics,* vol. 1. Edited by Nathaniel J. Mass. Cambridge, Mass.: MIT Press.

Schweppe, Fred C. 1973. *Uncertain Dynamic Systems.* Englewood Cliffs, N.J.: Prentice-Hall.

Senge, Peter M. 1975. "Future Electronics Company." System dynamics group working paper D–2310, MIT, Cambridge, Mass.

Senge, Peter M. 1978. "The System Dynamics National Model Investment Function: A Comparison to the Neoclassical Investment Function." Ph.D. dissertation, MIT, Cambridge, Mass.

Theil, Henri. 1971. *Principles of Econometrics.* New York: John Wiley and Sons.

Turoff, M. 1970. "The Design of a Policy Delphi." *Technological Forecasting and Social Change,* 2:149.

Appendix: Model Equations

```
R       HD.K=H.K/HL
C       HL=66
R       HC.K=H.K*HCN*HLM.K
C       HCN=0.07
A       HLM.K=TABLE(HLMT,LFO.K,0,1,.1)
T       HLMT=.8/.95/1.075/1.2/1.3/1.35/1.35/1.25/1/.6/0
A       LFO.K=(H.K*LPH)/AREA
C       LPH=0.1
C       AREA=9000
L       H.K=H.J+(DT)(HC.JK-HD.JK)
N       H=HN
C       HN=14000
NOTE
NOTE    CONTROL STATEMENTS
NOTE
PLOT    H=H(0,*)/HC=C,HD=D(0,*)
C       DT=0.5
C       LENGTH=80
C       PLTPER=2
RUN
```

Estimating Lengths and Orders of Delays in System Dynamics Models

8

Margaret S. Hamilton

8.1 Introduction

Delays are a ubiquitous feature of dynamic systems; they are present at every stage of an action. Time is required to recognize a problem, to decide what to do about it, and to implement a decision once made. Policy-makers must understand delays if they are to foresee the consequences of their actions. Many decisions turn out to be faulty because people underestimate the length of delays. Wage-price controls are adopted and then abandoned as ineffectual before they have had the desired effect. Monetary policy is changed before results can be seen. It is often not sufficient to ask an "expert" how long it will take for the repercussions of an action to be complete, because even the "experts" can seriously underestimate delay times. For example, in 1971 collective wisdom based on intuition put the time that it would take for most of the effects of the Smithsonian currency realignment to work through at one and one-half to two years. However, empirical research has shown that the time required for such changes is closer to five years (Junz and Rhomberg, 1973). The "experts" were off by a factor of 3. It is therefore important to have systematic methods of estimating adjustment times in dynamic systems.

The time structure of delays must also be understood by policy-makers. Whether a delay is destabilizing or stabilizing will depend on whether the repercussions are concentrated or dispersed, as well as whether the time lag is long or short. For example, governments often try to stimulate investment in a counter-cyclical fashion. If there is a long lag between changes in policy instruments and changes in actual expenditure, and if the final results are concentrated in time, policies designed to stimulate investment may be destabilizing instead of stabilizing. Under such circumstances the forecasting of economic conditions and the timing of policy measures will need to be very precise, and no policy at all may be preferable to one that does not take these considerations into account. It is thus also important to have systematic methods of estimating the time structure of delays.

Not all lengths and orders of delays in a model need to be estimated. It is often the case that system dynamics models are completely insensitive to the order of a delay and occasionally the case that even the length is unimportant. Sensitivity analysis should be employed to determine the crucial parameters before substantial effort is expended on estimation. Once the sensitive parameters have been determined, one of the statistical techniques presented here can be used for estimation purposes.

8.2 Types of Delays

Length and Order of Delays

When an action occurs, its consequences are felt over a period of time. A delay can be described by the time path of the repercussions. Characteristics to be observed of a time path can be its total length (the time required for the effects of an action to be fully worked out), its average length (essentially the half-way point, or the time required for one-half of the consequences to be felt),[1] and its shape (whether it is dispersed or concentrated, whether effects are felt immediately or only after some time has elapsed).

Delays in system dynamics models are defined by two parameters: average length and order (total length is always infinite). The length of a delay, or adjustment time, *AT*, is the average length, and the order of a delay is an integer greater than or equal to 1 that determines the shape of the delay. A first-order delay of a variable produces an exponential average of past values of that variable (weighting more recent values more heavily); an *n*th-order delay is a sequence of *n* first-order delays. Consequently, low-order delays have an immediate response that is dispersed; high-order delays have a deferred response that is more concentrated).[2] By varying the length and order of a system dynamics delay, we can approximate most well-behaved time paths. Figure 8.1 illustrates this variation. The first three delays have the same length but are of different orders. The three delays in the second row are of the same order as those immediately above them but have longer lengths.

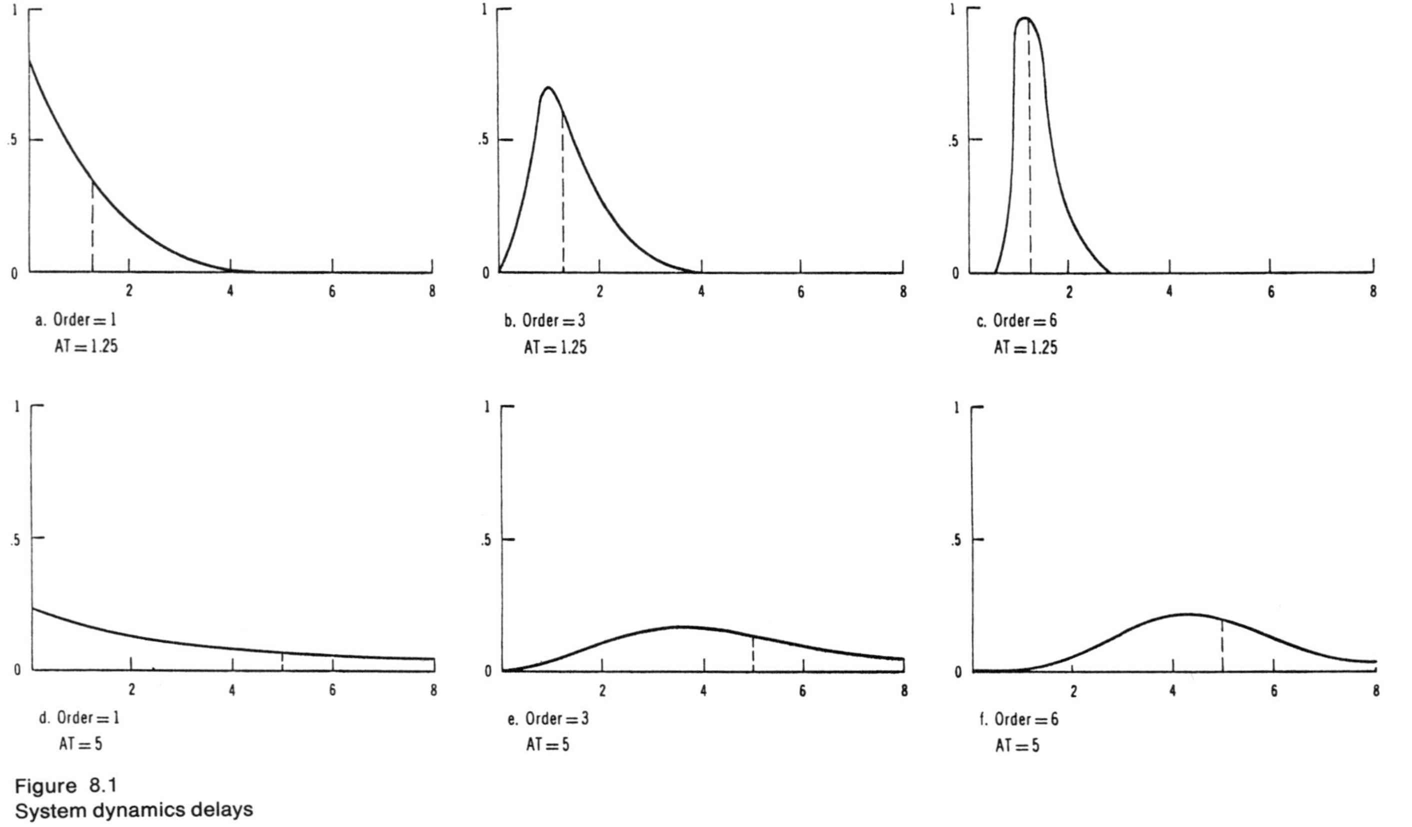

Figure 8.1
System dynamics delays

Material and Information Delays

Delays can be subdivided into material and information delays. A material delay modifies a physical flow such as letters in the mail. If one hundred letters are mailed, they are not immediately received. Their arrival will be dispersed over a period of time, the average length of which will depend on postal conditions.

By contrast, information delays are delays in perception or reaction. People may not respond to changed conditions unless they feel that the change is permanent. A temporary increase in the price of a commodity will not cause producers to increase the supply of that commodity. If, however, the increase persists over a period of time, the change will be perceived as permanent and increased production will be initiated.

Most real-world delays combine information and material delays. For example, in making an investment expenditure, time is required to collect information about market conditions and to reach a decision on the basis of this information. This period of time is an information delay. When the order for capital equipment has been placed, time is required for the production, delivery, and installation of the equipment. This period of time is a material delay. Since the time paths of material and information delays do not differ very much, there is no need to distinguish between the two types in estimating their lengths and orders.

8.3 General Formulation of the Problem

To formulate the concept of a delay in terms that facilitate estimation, it is useful to consider an example. Suppose that we mail X_t letters in time period t and receive, on the average, w_1X_t of those letters in period $t+1$, w_2X_t in period $t+2$, and so forth, up to w_mX_t in period $t+m$ (where $\Sigma w_i = 1$). Then, if the system of reactions is constant over time, X^*_t, the expected number of letters received in any period t can be written as a linear function of the m previous values of X_t:

$$X^*_t = w_1X_{t-1} + w_2X_{t-2} + \dots + w_mX_{t-m}, \qquad (8.1)$$

$$w_i \geqslant 0, \qquad i = 1, \dots, m,$$

$$\sum_{i=1}^{m} w_i = 1.$$

Equation (8.1) is quite general. X_t^* could be the expected price of a commodity in period t, determined by a weighted average of prices in previous periods, or it could be the expected income of an individual, determined by a weighted average of past incomes. The numerical values of the weights w_i will depend on the postal system in the first example; they will depend on the way in which expectations are formed in the second of the two cases.

Expressed in this fashion, a delay can be thought of as a probability distribution; w_i is the probability of a discrete random variable taking on the values 1, 2, . . . , m. In the postal example, w_i is the probability that a letter mailed in period t will arrive in period $t + i$. The adjustment time, AT, or average length of the delay, is the mean of the probability distribution:

$$AT = \sum_{i=1}^{m} iw_i . \tag{8.2}$$

In other words, AT is the average length of time a letter spends in the mail. The problem is to estimate the weights w_i from data.

More generally, we may wish to estimate Y_t, assumed to be a linear function of X_t^*:

$$Y_t = a + bX_t^* + u_t, \qquad u_t \sim N(0,\sigma^2), \tag{8.3}$$

where u_t is a random disturbance. Substituting equation (8.1) into equation (8.3) we obtain

$$Y_t = a + b(w_1X_{t-1} + w_2X_{t-2} + \ . \ . \ . + w_mX_{t-m}) + u_t, \tag{8.4}$$

$$Y_t = a + b_1X_{t-1} + b_2X_{t-2} + \ . \ . \ . + b_mX_{t-m} + u_t, \tag{8.5}$$

where $b_i = w_ib$
$i = 1, \ldots, m.$

AT can then be written

$$AT = \sum_{i=1}^{m} iw_i = \sum_{i=1}^{m} (ib_i / \sum_{i=1}^{m} b_i). \tag{8.6}$$

If we know the coefficients b_i in equation (8.5), we can calculate the adjustment time form equation (8.6). Plotting the weights w_i against lagged time gives us the time structure of the delay.

8.4 Estimation Techniques

Direct Estimation

In some cases it is possible to use ordinary least squares to estimate equation (8.5) directly. However, problems often arise. If the length of the lag (the number of periods over which the effect of an action is distributed) is long, there will be a large number of parameters to estimate $(m + 1)$. There may not be sufficient degrees of freedom to perform the estimation.

Another serious problem is multicollinearity. Frequently the various lagged values of X are highly intercorrelated, leading to imprecise estimates of their coefficients (their variances and covariances are large). If the estimates of the coefficients are poor, estimates of the length and order of the delay will be unreliable as well. Multicollinearity is most apt to occur with time-series data when the period of observation is short.

In many estimation problems, detecting multicollinearity can be quite difficult. Fortunately this is not likely to be the case for direct estimates of time structures. Since the regressors are lagged values of the independent variable X, we should not expect collinearity to occur in a very complicated form, and examining the simple correlation coefficients between the lagged variables should be sufficient to test for multicollinearity.

Prespecifying the Coefficients

Weymar (1968) tried to use the direct technique to estimate delays in his study of the dynamics of the world cocoa market. While working with a small number of observations of monthly time-series data, he encountered severe problems with multicollinearity and degrees of freedom. To circumvent these problems, he specified a priori two lag distributions (shapes of time paths), one wide and one narrow, and shifted these distributions along the time axis to test different lag times. This procedure amounts to prespecifying the coefficients w_i instead of estimating them. He chose the distribution and average lag that explained the largest amount of variance in cocoa consumption. Figure 8.2 shows Weymar's wide and narrow distributions for monthly cocoa consumption. He tested each distribution with several average lag times and arrived at a mild preference for the wide distribution with an average lag time of seven months.

Weymar's procedure is very cumbersome and, unless the length and order of the delay are fairly well known in advance, involves performing a very large number of regressions. Several other methods of avoiding the problems associated with the direct approach are available. These methods consist of specifying a priori some assumption about the form of the weights but not numerical values for them.

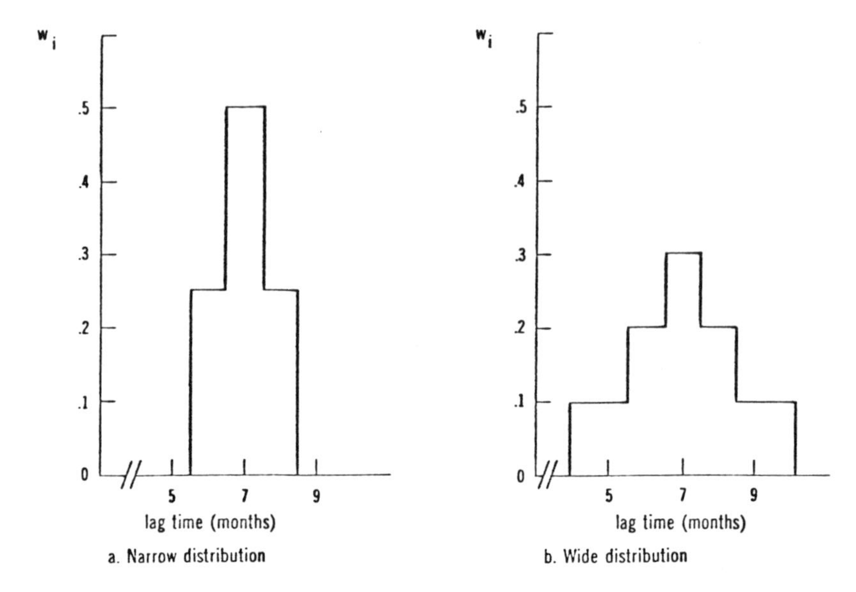

Figure 8.2
Weymar's structures for monthly cocoa consumption

Geometric Lag Distribution

If there is reason to believe that a delay is first order, a geometric lag can be used to estimate its length.[3] In a geometric lag model, the weights w_i decline exponentially:

$$w_i = (1 - \lambda)\lambda^{i-1}, \qquad 0 < \lambda < 1, \qquad i = 1, 2, \ldots \tag{8.7}$$

Here, the most recent values of X are given the largest weights. The effect of X on Y extends indefinitely into the past ($m = \infty$), but the weights decline in a fixed proportion so that the effect of distant values of X eventually becomes negligible. Substituting equation (8.7) into equation (8.4), we obtain

$$Y_t = a + b(1 - \lambda)(X_{t-1} + \lambda X_{t-2} + \lambda^2 X_{t-3} + \ \ldots) + u_t. \tag{8.8}$$

Equation (8.8) can not be estimated because it involves an infinite number of regressors. However, Koyck (1954) showed how to simplify the equation as follows: lag equation (8.8) one period, multiply through by λ, and subtract the result from the original equation to obtain

$$Y_t = a(1 - \lambda) + b(1 - \lambda)X_{t-1} + \lambda Y_{t-1} + v_t, \tag{8.9}$$

$$v_t = u_t - \lambda u_{t-1}.$$

Equation (8.9) contains only three parameters to estimate—a, b, and λ—and two regressors—Y_{t-1} and X_{t-1}. There should be no problem with degrees of freedom and multicollinearity.

Equation (8.7) defines a geometric probability distribution with mean $1/(1-\lambda)$ implying that

$$AT = 1/(1-\lambda). \tag{8.10}$$

The adjustment time can be calculated from (8.10) using $\hat{\lambda}$, the estimated coefficient of Y_{t-1} in equation (8.9), as an approximation to λ.

There are two drawbacks to using geometric lags to estimate adjustment times. First, it is necessary to assume a priori that the delay is first order, but it is preferable to let the data determine the order of the delay. Second, and more serious, it is not possible to use ordinary least squares to estimate the coefficients a, b, and λ in equation (8.9) because the disturbance term v_t is correlated with Y_{t-1}, one of the explanatory variables. In fact,

$$\begin{aligned} \mathrm{E}[v_t Y_{t-1}] &= E[u_t - \lambda u_{t-1})(a + b(1-\lambda)(X_{t-2} + \lambda X_{t-3} + \ldots) + u_{t-1})] \\ &= -\lambda\sigma^2 \end{aligned}$$

where σ^2 is the variance of u_t. Ordinary least-squares estimates of the coefficients will be inconsistent (they will be biased, and the bias will persist even as the number of observations approaches infinity). Some other estimation technique, such as the method of instrumental variables, must be used.[4]

In system dynamics models, higher-order delays are obtained by aggregating several sequential first-order delays (cascading several levels.) The probability distribution for an nth-order delay can be obtained by convoluting (summing sequentially) n independent and identically distributed geometric random variables. The resulting random variable has a negative binomial distribution:

$$w_i = \left(\frac{i+n-1}{i} \right)(1-\lambda)^n \lambda^i, \quad i = 0, 1, \ldots, \tag{8.11}$$

with mean $n\lambda/(1-\lambda)$. A distributed lag model with weights defined by equation (8.11) is known as a Pascal lag model. Such a model can be estimated and the adjustment time determined from the estimated coefficients, but the procedure is not recommended. Pascal models are difficult to estimate and, in addition, have all the problems associated with the geometric distributed lag, that is, the order must be specified in advance and the disturbance terms are serially correlated. Other methods of estimating higher-order delays exist which, while not mathematically equivalent to higher-order system dynamics delays, are easy to use and give similar results.

Polynomial Lag Distribution

A polynomial lag, suggested by Almon (1965), is a very flexible approach and can be used to estimate both the length and the order of a system dynamics delay. The idea is to approximate the time path with a polynomial. The method is based on Weierstrass's theorem, which states that a function, continuous in a closed interval, can be approximated with any prespecified accuracy by a polynomial of suitable degree. Since most time paths are reasonably well behaved, a polynomial of fairly low degree (3 or 4) should suffice. The degree n of the polynomial and the total length of the lag k must be chosen in advance. These choices can be tested by varying both parameters. The weights w_i will lie along a polynomial of the chosen degree:

$$w_i = \lambda_0 + \lambda_1 i + \lambda_2 i^2 + \ldots + \lambda_n i^n, \quad i = 1, \ldots, k. \tag{8.12}$$

The number of parameters to be estimated depends only on the degree of the polynomial, not on the length of the lag. Therefore, if the degree is fairly low, there should be no problem with degrees of freedom. Since the explanatory variables involve different powers of i, multicollinearity should not be a problem. The original disturbance term appears as the disturbance term in the transformed equation; so, if there was no serial correlation to begin with, none has been introduced. As a result ordinary least squares can be used to perform the estimation, and the resulting estimates should have all the desirable properties. Standard econometric texts discuss how to transform the original equation into a suitable one for estimation, as well as how to reduce the number of parameters that must be estimated (by constraining the polynomial to cross the time axis at particular points and forcing the weights to sum to 1). For example, see Kmenta (1971, pp. 492–493).

Once the weights have been estimated, the adjustment time can be calculated from equation (8.6). The weights w_i can be plotted against lagged time and compared to computer plots of various-order system dynamics delays with the calculated adjustment time. This procedure should give a fair estimate of the order of the delay. The modeler should not try to match the order too accurately, since there is no exact equivalence between an nth-order system dynamics delay and a polynomial lag distribution. It should be sufficient to determine if the delay is of first, third, or higher order.

Fitting Erlang Distributions

The estimation techniques discussed so far have all been forms of distributed lags. Using these techniques requires time-series data on the dependent and independent variables Y and X (the letter receiving rate and letter-mailing rate, for example). It is not necessary to be able to associate particular letters received with particular letters mailed. In some cases, however, additional data may be available. Consider an example: if the system to be modeled is a hospital, the hospital may keep records of the number of patients that stay less than one week, between one and two weeks, and so forth, as well as records of the number of patients being admitted and discharged each week. When this sort of detailed information is available, there is a particularly simple method of calculating the length and order of the delay (the delay being the stay in the hospital.) The method is based on the fact that the continuous probability distribution associated with a kth-order system dynamics delay is Erlang type k.

The probability distributions considered so far (the geometric and negative binomial distributions corresponding to first- and higher-order delays) have been discrete, of necessity, because the data were discrete. A system dynamics model, however, is continuous. The true probability distribution associated with a first-order delay is the exponential distribution:

$$(1-\lambda)e^{-(1-\lambda)t}, \qquad t > 0. \tag{8.13}$$

This is the limiting case of the geometric distribution as the period of observation (time increment) becomes smaller and smaller. The distribution associated with a kth-order delay can be obtained by convoluting (summing sequentially) k independent and identically distributed exponential random variables. The resulting random variable has an Erlang type k distribution:

$$\frac{[k(1-\lambda)]^k}{(k-1)!}\, t^{k-1} e^{-k(1-\lambda)t}, \qquad t > 0. \tag{8.14}$$

The exponential distribution is the special case of the Erlang obtained by setting $k = 1$. The Erlang type k distribution has mean $1/(1-\lambda)$ and variance $1/[k(1 - \lambda)^2]$. Figure 8.3 shows Erlang distributions with various values of k. As $k \to \infty$, the distribution becomes completely deterministic (that is, it becomes concentrated at the mean). This outcome can also be seen from the formula for the variance; the variance becomes 0 as $k \to \infty$.

The formulas for the mean and variance of an Erlang distribution can be used to calculate the length and order of the delay. Let n_i be the number of patients staying in the hospital from $i - 1$ to i weeks, $i = 1, 2, \ldots$ The sample mean is

$$\bar{X} = \frac{\sum_i i n_i}{\sum_i n_i} \doteq \frac{1}{(1 - \lambda)} = AT. \tag{8.15}$$

The adjustment time AT is approximately equal to the sample mean. The sample variance is

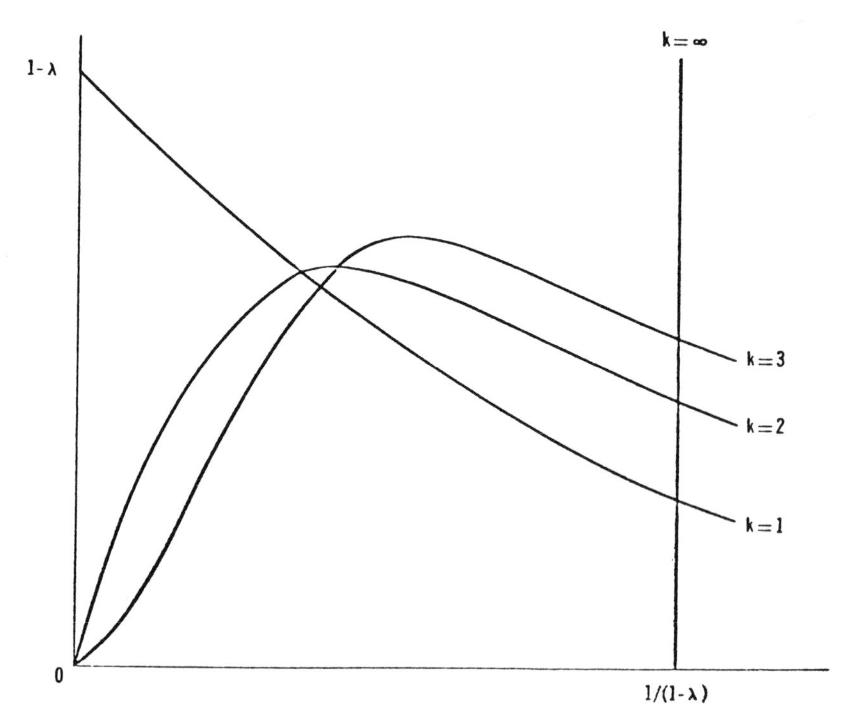

Figure 8.3
A family of Erlang distributions with mean $1/(1 - \lambda)$

$$s^2 = \frac{\sum_i n_i (i - \bar{X})^2}{\sum_i n_i} \doteq \frac{1}{k} \frac{1}{(1-\lambda)^2} = \frac{1}{k} \bar{X}^2,$$

implying that

$$\hat{k} = \bar{X}^2 / s^2, \tag{8.16}$$

where $\hat{k}$, an estimate of the order k, is equal to the square of the sample mean divided by the sample variance. Of course, $\hat{k}$ won't necessarily be an integer, but if $\hat{k} = 2.85$, for example, we can conclude that the delay is third-order. This calculation is so simple (it involves no regression) that it should be used whenever the appropriate data are available.

8.5 Empirical Estimations

An Empirical Study Using the Direct Method

Junz and Rhomberg (1973) used the direct approach to estimate the timing of the effects of relative price changes on export flows. Such changes in relative prices can be brought about either by alterations in exchange rates or by changes in export prices measured in national currencies. Proportional changes in market shares were related to proportional changes in relative export prices, yielding a price elasticity of market shares. Such a study aggregates several information and material delays: the recognition lag required for the changed competitive situation to be perceived, the replacement lag required for inventories to be depleted or equipment to wear out, the production lag required for producers to undertake the expense of shifting from supplying one market to supplying another or of adding capacity in order to supply additional markets, to name only a few. Whether it is advisable to model (and therefore estimate) each delay separately or combine them in one aggregate delay depends on the model being constructed and whether disaggregation changes the dynamic behavior of the system. In what follows, all of the constituent delays have been combined.

The data consist of annual observations for the years 1958 to 1969 (allowing for five lags in price data going back to 1953) for thirteen exporting countries and thirteen markets in each country. The fitted equation is

$$PCMS_t = 0.019 - \underset{(2.2)}{0.52PCP_t} - \underset{(1.2)}{0.29PCP_{t-1}} - \underset{(2.4)}{0.58PCP_{t-2}}$$
$$- \underset{(4.2)}{0.98PCP_{t-3}} - \underset{(1.0)}{0.24PCP_{t-4}} - \underset{(1.2)}{0.27PCP_{t-5}},$$

where $PCMS$ stands for percent change in market share and PCP for percent change in relative export price. The numbers shown in parentheses are t ratios. Three of the six regression coefficients are statistically significant at the 95 percent confidence level.

The response can be divided into an immediate response (for items that are produced quickly and are relatively homogeneous across suppliers) and a delayed response. Figure 8.4a shows the delayed response plotted against lagged time in years. A smooth curve has been drawn to

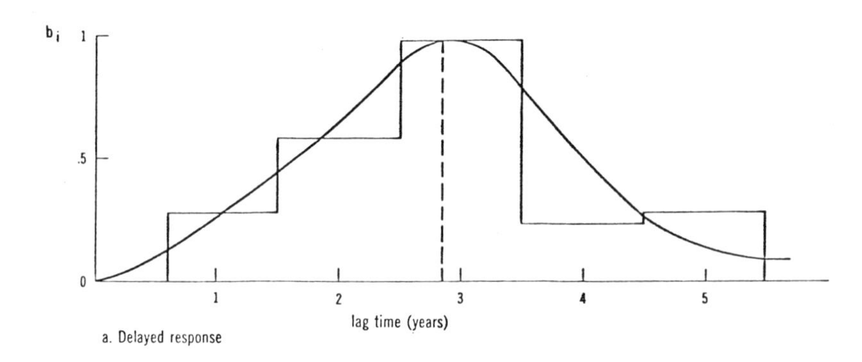

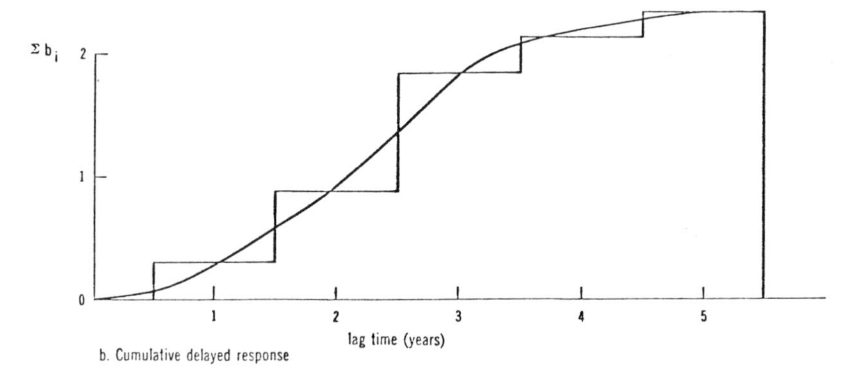

Figure 8.4
Time structures estimated by Junz and Rhomberg (1973)

approximate the time structure. It can be seen that in the fifth year there is still quite a lot of adjustment taking place. Figure 8.4b shows the cumulative delayed response (the sum of the responses up to any point in time) plotted against time. The adjustment time can be calculated from equation (8.6):

$$AT = \sum_{i=1}^{5} ib_i / \sum_{i=1}^{5} b_i = 2.84 \text{ years.}$$

The peak response occurs in the third year.

Figures 8.5a and 8.5b show pulse and cumulative responses of third-order system dynamics delays with adjustment times of 2.84 years. Figures 8.5c and 8.5d give similar plots for sixth-order delays. The sixth-order delay gives a closer approximation to the estimated time path, but not too much weight should be given to the exact order of the delay. What is important is that it is of higher order.

An Empirical Study Using Polynomial Lags

Almon (1965) employed polynomial lags to predict quarterly capital expenditures in manufacturing industries from present and past capital appropriations. The quarterly data on appropriations and expenditures came from the survey conducted by the National Industrial Conference Board among the thousand largest manufacturing companies in the United States. Estimates were based on the nine years, from 1953 to 1961. Dummy variables were added to the equation to remove the effects of seasonal variation in expenditures. The length of the lag was assumed to be seven years. The resulting equation for all manufacturing industries (neglecting the seasonal variables and the constant term) is

$$\begin{aligned} KE_t = {} & \underset{(2.09)}{0.048KA_t} + \underset{(6.19)}{0.099KA_{t-1}} + \underset{(10.85)}{0.141KA_{t-2}} + \underset{(7.17)}{0.165KA_{t-3}} \\ & + \underset{(7.26)}{0.167KA_{t-4}} + \underset{(11.23)}{0.146KA_{t-5}} + \underset{(6.56)}{0.105KA_{t-6}} + \underset{(2.21)}{0.053KA_{t-7}}, \end{aligned}$$

where KE and KA stand for capital expenditures and appropriations, respectively. The numbers shown in parentheses are t ratios. All of the lagged variables are very significant. The weights, which were not constrained to sum to 1, add up to 0.924. The difference between 0.924 and 1 can be very nearly accounted for by cancellations. The adjustment time[5] can be calculated as

$$AT = \sum_{i=0}^{7} ib_i / \sum_{i=0}^{7} b_i = 3.55.$$

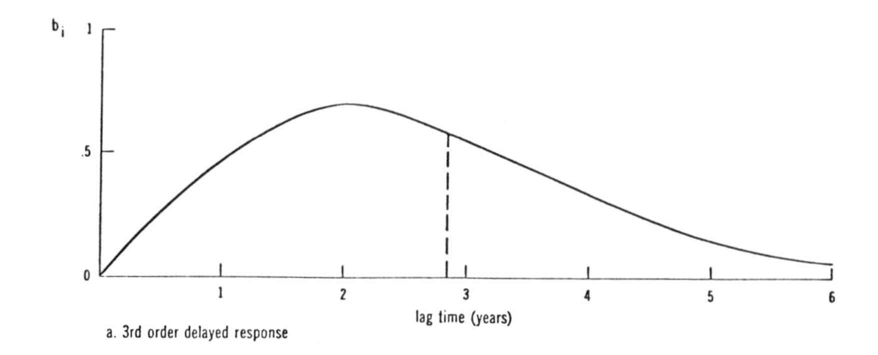

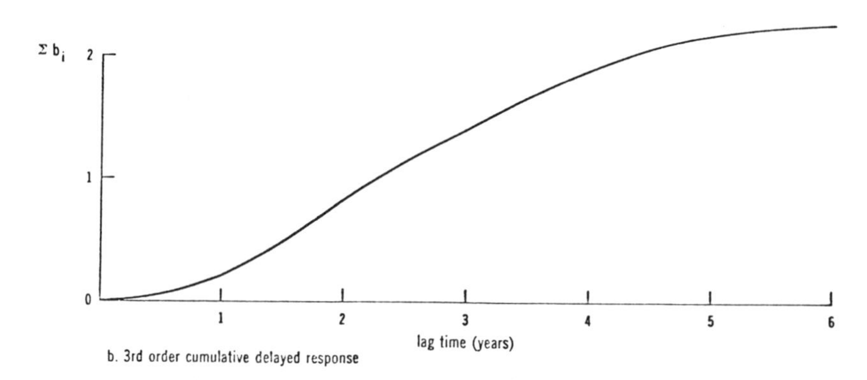

Figure 8.5
System dynamics third- and sixth-order delays with adjustment times of 2.84 years

The peak response comes in the fourth year. Figures 8.6a and 8.6b show the pulse and cumulative responses of the estimated time paths; figure 8.7a and 8.7b show similar responses for third-order system dynamics delays with the same adjustment time.

8.6. Choosing an Estimation Technique

Table 8.1 summarizes the techniques for estimating lengths and orders of delays in system dynamics models and the advantages and disadvantages of each.

Having decided through sensitivity analysis that a delay should be

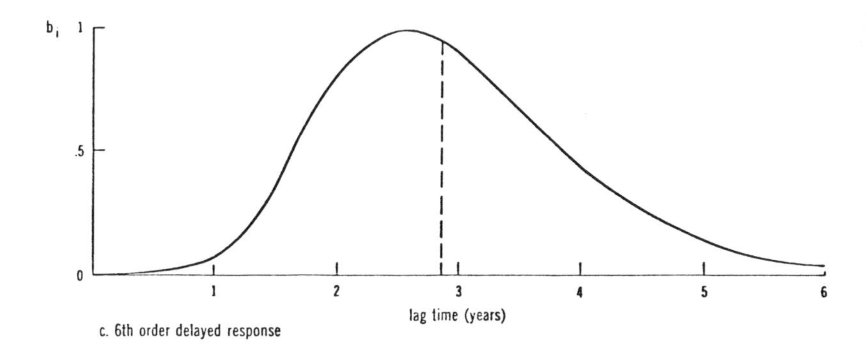

c. 6th order delayed response

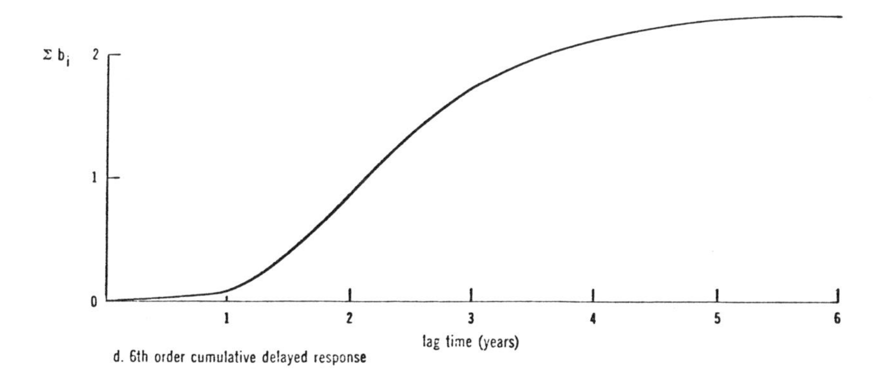

d. 6th order cumulative delayed response

estimated, the modeler is faced with a choice among five techniques. Here are a few guidelines:

1. If the data required to fit an Erlang distribution are available, this method should be used; it is by far the simplest approach and is exactly equivalent to a system dynamics delay.
2. If there are sufficient observations, and multicollinearity does not present a problem, then the direct technique should be used.
3. If there is reason to believe that a delay is first-order, or if the order of the delay is not important, and if the modeler has access to a program that corrects for serial correlation, then a geometric lag should be used.
4. The polynomial lag, though somewhat more cumbersome, provides the greatest flexibility and overcomes most of the special conditions that need to be met before the other methods can be used. This technique will work in nearly all circumstances.

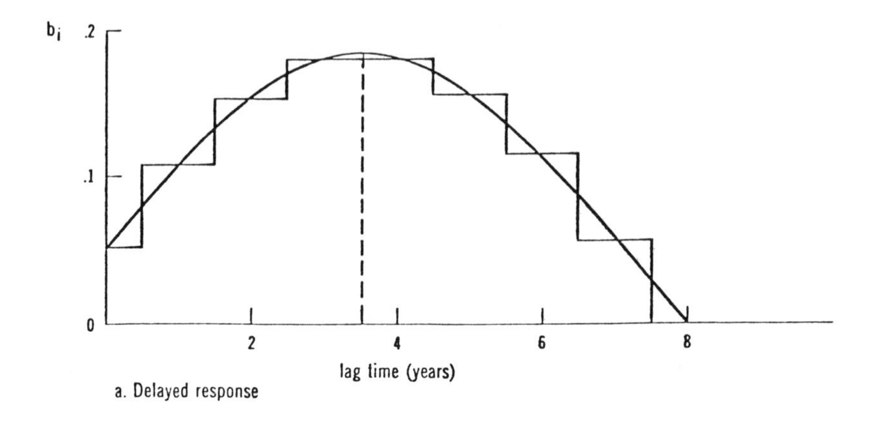

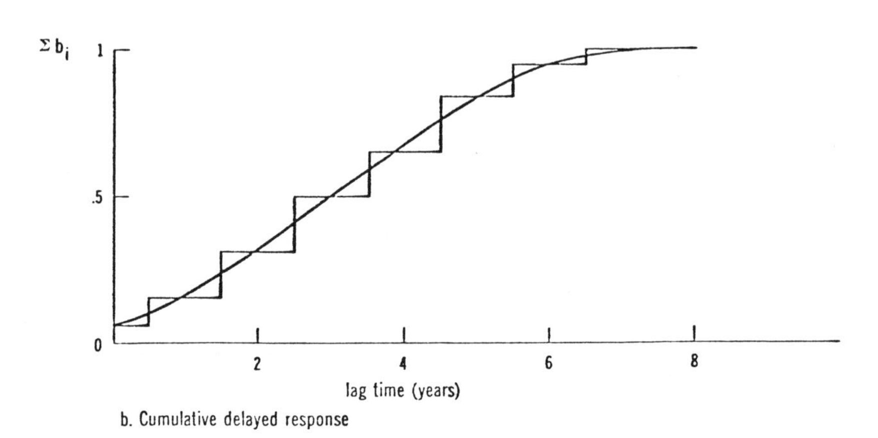

Figure 8.6
Time structures estimated by Almon (1965)

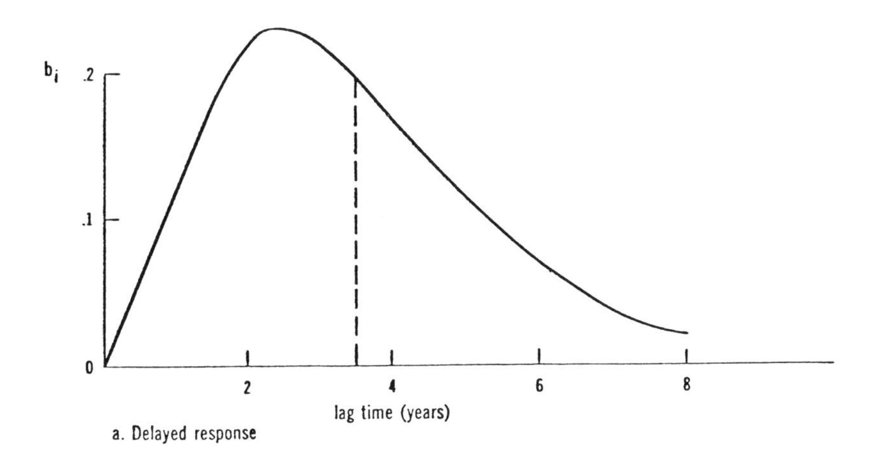

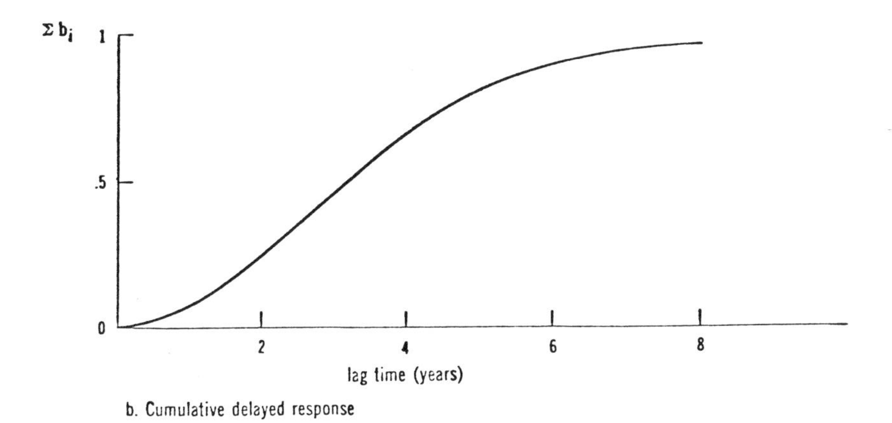

Figure 8.7
System dynamics third-order delays with adjustment times of 3.55 years

Table 8.1

Method	Advantages	Disadvantages
Direct	Uses observations directly	There may be too many parameters to estimate (too few degrees of freedom)
	No need to specify the order of delay	The explanatory variables may be highly intercorrelated (multicollinearity)
Prespecified coefficients	Uses observations directly	Length of delay must be specified
		Shape of delay must be specified
		Many regressions may have to be performed
Geometric lag	Uses observations directly	Useful only for first-order delays
	Simple functional form	Serial correlation of residuals (ordinary least squares can not be used)
	Exact equivalence with first-order delay	
Polynomial lag	No need to specify order of the delay	Variables must be transformed before estimation can take place
Fitting an Erlang distribution	Very simple calculation	Requires additional data
	Exact equivalence with system dynamics delay	
	No need to specify the order of the delay	

8.7 Validation

Once estimates have been obtained, how does the modeler develop confidence in them? There are several things that can be done to test the robustness of the estimates:

1. The length of the lag can be varied for either the direct or the polynomial method; the degree of the polynomial can be varied for the latter.
2. The data can be disaggregated and new estimates made. For example, in the estimation of market shares, single markets can be considered; in the estimation of capital expenditures, manufacturing can be separated into durables and nondurables or even further disaggregated.
3. As new data become available, they can be added to the old, and new estimates can be made.

If the resulting time structures are similar to the original ones, confidence in the estimates will be improved.

The two applications described in this chapter differ in one fundamental respect. Whereas nearly all capital expenditures can be accounted for by appropriations minus cancellations, variation of relative prices is only one of the determinants of export market shares. In the latter case, variables not included in the model are thrown into the residual category. It would therefore be of interest to know just how sensitive the estimates are to noise. Sensitivity to noise can be tested by adding a stochastic variable that might be correlated with the calculated residuals of the regression equation. If the estimates of the lag coefficients are not much changed by such an addition, confidence in them will be further increased.

Notes

1. Strictly speaking, the average length of a delay in a system dynamics model is not the abscissa of the vertical line that divides the area under the time path into two equal parts; it is the abscissa of the center of gravity of the area under this curve.

2. For a more complete description of system dynamics delays, see Forrester, 1968.

3. The DYNAMO rate equation of a first-order delay macro can be expanded to show that it is precisely equivalent to a geometric lag:

DYNAMO equations

$R \quad O \cdot KL = L \cdot K/AT$
$L \quad L \cdot K = L \cdot J + (DT)(I \cdot JK - O \cdot JK)$

Equations with t subscripts (assuming $DT = 1$)

$$O_t = L_t/AT$$
$$L_t = L_{t-1} + I_{t-1} - O_{t-1}$$
$$O_t = L_t/AT$$
$$= (L_{t-1} + I_{t-1} - L_{t-1}/AT)/AT$$
$$= I_{t-1}/AT + \frac{(1 - \frac{1}{AT})}{AT} L_{t-1}$$
$$= \frac{I_{t-1}}{AT} + \frac{(1 - \frac{1}{AT})}{AT} I_{t-2} + \frac{(1 - \frac{1}{AT})^2}{AT} I_{t-3} + \ldots.$$

From equation (8.10), $AT = \frac{1}{(1 - \lambda)}$ and $\lambda = 1 - (\frac{1}{AT})$. Substituting, we obtain

$$O_t = (1 - \lambda)I_{t-1} + (1 - \lambda)\lambda I_{t-2} + (1 - \lambda)\lambda^2 I_{t-3} + \ldots.$$

The output O_t is a geometric lag of the input I_t.

4. A description of the method of instrumental variables can be found in any standard econometrics text, such as Kmenta, 1971, pp. 479–480. For a discussion of the problems of estimation with serially correlated disturbances, see Griliches, 1967.

5. The subscript i runs from 0 to 7 in this example because Almon chose to include the present value of X in the lag structure.

References

Almon, Shirley. 1965. "The Distributed Lag Between Capital Appropriations and Expenditures." *Econometrica,* 33 (January), pp. 178–196.

Forrester, Jay W. 1968. *Principles of Systems. Cambridge, Mass.: Wright-Allen Press.*

Griliches, Zvi. 1965. "Distributed Lags, A Survey." *Econometrica,* 35 (January), pp. 16–48.

Junz, Helen B., and Rhomberg, Rudolf R. 1973. "Price Competitiveness in Export Trade Among Industrial Countries." *American Economic Review* (May), pp. 412–418.

Kmenta, Jan. 1971. *Elements of Econometrics.* New York: Macmillan Co.

Koyck, L. M. 1954. *Distributed Lags and Investment Analysis.* Amsterdam: North Holland Publishing Co.

Weymar, F. Helmut. 1968. *The Dynamics of the World Cocoa Market*. Cambridge, Mass.: MIT Press.

Part V
Testing

Sensitivity Analysis in System Dynamics

9

Carsten Tank-Nielsen

9.1 Introduction

Sensitivity analysis is an element of most formal modeling processes. However, as each field of modeling has its distinct and characteristic features, so has the sensitivity analysis that accompanies it. It is the purpose of this chapter to describe some of the central issues related to sensitivity analysis in system dynamics.

To portray the complete role of sensitivity analysis in system dynamics I will define sensitivity analysis as "the study of model responses to model changes." This is a rather broad definition, which should be kept in mind when reading the chapter.

Sensitivity analysis is a primary concern in system dynamics model-building and review. One reason is that problems commonly analyzed in system dynamics necessitate the incorporation of relationships and parameters for which little empirical data are available. Quantifying the system elements is often quite difficult. Consequently, any model evaluation must include a study of the effects of the uncertainty in the representation. Another reason is the complexity of the problems being modeled. The models are often hard to understand. Sensitivity analysis is one way of attaining increased understanding of how models work.

A complete description of sensitivity analysis must distinguish between

1. objectives in sensitivity analysis,
2. types of model change in sensitivity analysis,
3. the interpretation of model response to change,
4. efficient conduct of sensitivity analyses.

This chapter focuses on issues relating to the first three aspects. The fourth aspect is discussed by Sharp (1976) and to some extent by Graham (1976).

9.2 The Objectives

A sensitivity analysis should always be related to the purpose of the model under investigation. In this respect it is important to keep in mind not only the explicit goals of the model but also the goals inherent in the modeling technique being used.

The most common purpose for a system dynamics model is to explain the causes of an undesirable behavior mode and identify policy variables capable of eliminating the undesirable behavior. Implicit in the system dynamics discipline are also the objectives of finding the simplest recognizable structure that can explain the initial reference mode and of identifying those areas where further research is necessary and critical.[1] The following description of the objectives in sensitivity analysis is based upon a recognition of both the explicit and implicit goals of system dynamics models.

To Test the Effects of Uncertainties in Parameter Values

Uncertainties in a system dynamics model's parameter values may affect its response and thereby the conclusion derived from the model. As system dynamics models tend to (1) include parameters for which no observations exist and (2) analyze such a long time span that most quantities tend to be variables rather than parameters with stable values, uncertainties in model parameters are common. Typically, their values will be known within a range, but not precisely.

Testing alternative parameter values within the assumed range of uncertainty in order to evaluate the impact on model conclusions constitutes an important part of sensitivity analysis in system dynamics.

As system dynamics models normally will be insensitive to variations in most model parameters, provided that the variations are kept within a realistic range, the sensitivity testing should first aim at identifying those parameters to which the model is sensitive. Second, the sensitive parameters should be varied within their assumed range of uncertainty in order to see if that changes overall model behavior. If some parameter changes do change the behavior, more effort should be put into establishing their precise values. Another alternative may be to reformulate the model, so that it relies solely on better known parameters.

To Generate Insight

When talking about the capacity of a system dynamics model to generate insight, I am talking about two types of insight: first, about the relation between structure and behavior and, second, about increased understanding of the real world. Sensitivity analysis is an important and efficient tool to gain both types of insight. In the following I discuss the role of the sensitivity analysis in gaining these two types of insight.

Insights about Structure and Behavior

To attain an understanding of the relation between model structure and model parameters, on the one hand, and model behavior, on the other, is a necessity in system dynamics. This insight forms the basis for suggesting effective policies to deal with the problem at hand.

The goal of attaining model insight embraces a number of more detailed objectives:

1. To discover which behavior modes the model can generate.
2. To identify the model changes which can shift the model from one behavior mode to another. This identification helps to sort out the parameters and structural relationships whose precise values are of critical importance for model behavior, thereby establishing which aspects a more comprehensive study should focus on. Furthermore, the modeler discovers where to allocate limited resources. Finally, such identification helps to locate appropriate levers for efficient and robust policy.
3. To identify the active and dormant parts of the model structure. This procedure establishes a basis for finding the simplest recognizable structure that can generate the reference mode. To find such a simple model structure is often a goal in system dynamics modeling because it will indicate the most fundamental processes at work within the system. Moreover, as a forum for discussing the problem under study, a simple model is preferable.
4. To evaluate whether the dynamic behavior in models with exogenous inputs is generated by external or internal forces.

The ultimate objective of model understanding is to provide a basis for improved policy.

To gain such understanding, the model should be tested over an unusually wide range of changes in both parameters and structure. Only a very wide range can reveal the inherent dynamics of the model. Changes in structure typically imply that feedback loops are cut, so as to enable the modeler to find which parts of the structure contribute to the observed behavior of the model. One potential danger in this procedure is worth mentioning. Cutting a loop and observing the effect on model behavior may lead to the wrong conclusions. This is because the possible difference in behavior need not result from the loop that was cut but rather from the loop's interaction with other loops.

Insights about the Real World

The idea underlying formal modeling is that it may help us understand the real world. If the model matches the real-world system it is meant to portray, insights can be transferred to the real-world system. How then can we achieve sufficient confidence in a formal model to say that the model insights also are valid in the real world? Confidence can be built only be passing the model through a variety of tests. One test is to subject the model to comprehensive sensitivity testing. Test conditions should deliberately be defined so that the system operates under extreme situations, where it is often easier to reveal whether the model is plausible or not.

The result of the comprehensive sensitivity testing helps to answer the following questions related to model confidence:

1. Are the behavior modes produced by the model realistic?
2. Does the model's sensitivity (or robustness) accord with human knowledge of the real-world system?
3. Is the model (in)sensitive to the same perturbations as the real system?

If the answers are yes, the experimentation has helped to increase our confidence in the model. Otherwise, the sensitivity analysis will indicate some deficiency, either in the modeler's mental model, the formal model, or both. The deficiency necessitates a review of the deficient model.

Even if a comprehensive sensitivity testing of a model reveals no significant deficiency, we have, needless to say, no proof of the model's correctness. Testing can *increase* confidence in the model.

To Direct Further Work on Parameters and Structure

Lack of precise data and problems in determining which part of the available information is relevant have often been reasons for *not* making a formal model. In system dynamics we often take the opposite point of view. Making a rough model at an early stage and subjecting it to a sensitivity analysis can be useful in identifying what is important data and in distinguishing between relevant and irrelevant information. In this perspective, sensitivity analysis actually directs further work.

Ideally sensitivity testing will reveal what few parameters have the potential to alter the model's behavior mode. Effort should be put into estimating or reformulating these parameters, while the other parameters are left at their low level of precision—which still is sufficient to let the model fulfill its purpose.

Moreover, sensitivity testing ideally reveals what feedback loops govern the model's behavior. Consequently, further work should be directed toward verification and understanding of this part of the structure.

9.3 Types of Model Changes in Sensitivity Analysis

All changes, with the possible exception of perturbations in the exogenous variables (including noise functions), can be viewed as being either a structural or a parameter change, or a combination of the two. Altering the system boundary implies a structural change. Changing the initial value of a level is a parameter change. Reformulating part of the model may imply a structural as well as a parameter change.

Changes in model parameters and structure are usually very different in their impact on model behavior. The division line between them, however, is not always clear. One reason is that a parameter change may induce a structural change as well. Another reason is that parameters are in a way reduced structures. Underlying each parameter is a structure. Altering the value of a parameter reflects a change in the implicit structure generating the parameter.

The Parameter Change

By parameters I mean constants (including initial values and delay times) and table functions. Whether or not this category includes exogenous variables will not be discussed here (see Forrester, 1961, pp. 112–114 and 124–129).

System dynamics models are usually rather insensitive to parameter changes, as are real complex systems. The source of parameter insensitivity resides in the model structure, typically in the dynamic properties of the negative feedback loop. Such goal-seeking loops tend to counteract any alterations imposed by parameter changes. A second reason for parameter insensitivity is that model behavior is primarily generated by a few feedback loops. Changes outside these loops—in the nondominant or dormant loops—will normally not affect model behavior.

There are, however, exceptions. Most model structures are capable of exhibiting different behavior modes. Most models can be shifted from one mode to another by changes in the model parameters (and/or the exogenous variables).

Therefore, when saying that system dynamics models are insensitive to parameter changes, what actually is meant is that they are insensitive to most parameter changes. In most cases model behavior is sensitive to a few points in the system.

Considering the large number of parameters in most system dynamics models, an arbitrary search for the most sensitive parameters would be hopelessly time consuming. The "key parameters," however, are often found in or near positive feedback loops, delay times, and intersections of several positive and negative loops.

Parameter testing should focus on identification and change of "key parameters." In addition, parameters which are controversial, and thus disputable, should be tested, although not necessarily with the intention of altering model behavior. A disputable parameter's potential for altering behavior may be small. The purpose of the test will often be to demonstrate the unimportance of the precise parameter value with respect to model conclusions.

The Structural Change

A structural change in a system dynamics model is an alteration of a causal relationship in the model. A structural change will normally be visible in a causal loop diagram, whereas parameter changes can only be seen in model equations or in the table function graphs. As a rule system dynamics models are more sensitive to structural than parameter change. Different structures are normally capable of exhibiting different behavior modes.

Models are usually more sensitive to structural changes that affect positive loops than negative loops. A positive loop normally encourages exponential growth. A negative loop will be goal seeking, and consequently act to stabilize model behavior. The addition or removal of a positive loop may therefore have a substantial impact on model conclusions. On the other hand, adding a negative loop to a stable system may have little effect, unless the addition results in a structure with two interacting negative feedback loops. Such an alteration can cause system oscillations. Removal of a negative loop from a stable system may destabilize the system, provided that the loop was active in the structure. As with parameter variations, the impact of a structural change depends on its connection to the dominant part of the model structure. Changing a dormant structure has little effect.

One major objective in structural sensitivity analysis should be to discover what few basic mechanisms that underlie the observed model behavior. The other important objective is to evaluate the impact of controversial or disputable relationships.

Parameter Changes with Structural Implications

In some cases, the division line between parameter and structural change is unclear. As an example, I show below how changes in a table function induce alterations in model structures.

Consider the simple structure in figure 9.1, where A and B are assumed to be coupled through a table function. If the table is as shown to the left in figure 9.2, the structure consists of two negative loops.

Now, if the relationship between A and B were changed to the pattern in figure 9.3, this would obviously be a parameter change. However, the alterations would also imply a change in structure. Now the structure

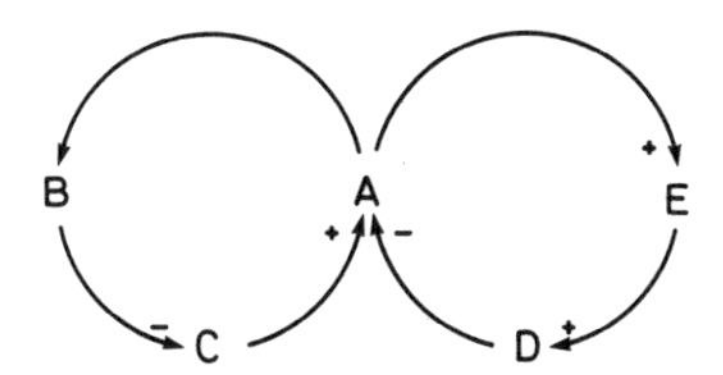

Figure 9.1
Model structure consisting of two coupled loops

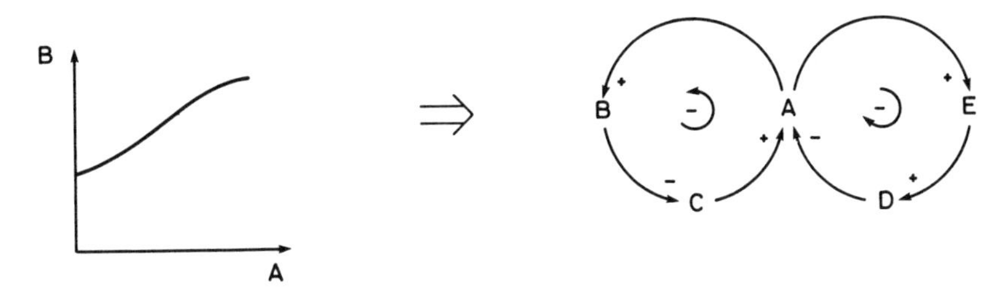

Figure 9.2
Positively sloping relationship between A and B resulting in two negative loops

consists of one positive and one negative loop, instead of two negative ones.

The change is much more than a parameter change. Further, the relationship between A and B could be as in figure 9.4. Once more, the parameter change has structural implications. The loop to the left in the figure will be positive or negative, depending on the value of A within its range.

Finally, changing the relationship between A and B, as shown in figure 9.5, eliminates one of the two loops completely. Here B is constant, regardless of the value of A, and the "loop" containing A, B, and C is inactive.

Reasonable versus Unreasonable Changes

In evaluating the effects of uncertainty in model parameters, changes should be restricted to values within the assumed range of uncertainty. Alterations which go beyond this range will not reflect the model's sensitivity to uncertainties in parameters. There are, however, some further requirements that must be satisfied in order for a parameter change to be reasonable.

System dynamics models often trace the behavior of a system through a historical period and into the future. A change in parameters that

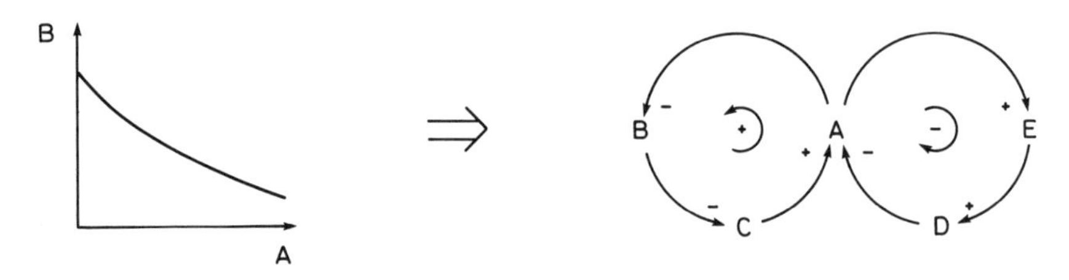

Figure 9.3
Negatively sloping relationship between A and B resulting in one positive and one negative loop

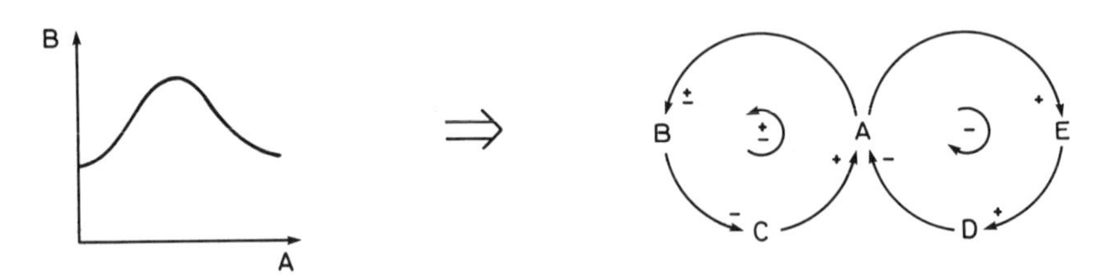

Figure 9.4
A bell-shaped relationship between A and B and its structural consequence

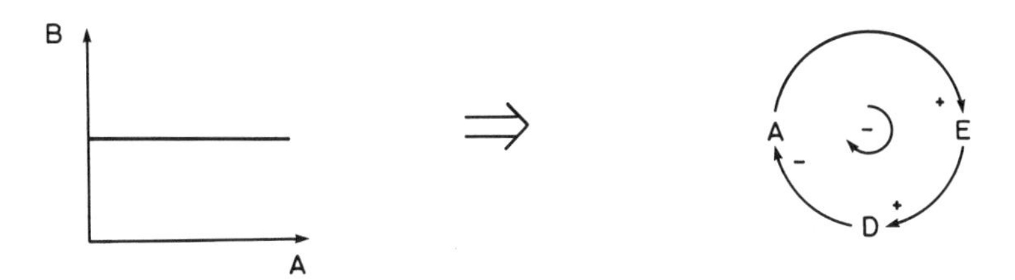

Figure 9.5
A flat table function and its structural consequence

destroys the historical fit of the model should not be viewed as a reasonable change. On the contrary, the change itself should be viewed as inconsistent with the rest of the model. In such cases adjustments should be made simultaneously in other parameters, so that the historical fit is maintained. This does not preclude sensitivity testing but restricts the range of plausible parameter sets. Taken together, the two or more changes may still recreate history, while predicting an alternative future. If such simultaneous adjustments prove impossible, even for a reasonable initial test value, one may suspect the model.

While striving to attain model understanding, however, the concept of "a reasonable change" should be interpreted somewhat differently. Changes need no longer be restricted to realistic values. To attain model understanding implies discovering the possible behavior modes of the system and finding the values at which the shift from one mode to another occurs. Therefore nearly any change would be considered reasonable.

In testing how well the model duplicates the real world, reasonable changes include any changes that are meant to reflect a conceivable real-world event. In such testing, variations in parameters should be restricted to values that they could realistically assume. Structural changes should always have a real-world counterpart. It may seem that a real-world occurrence can be represented by a change in a single parameter, but commonly such events correspond to simultaneous changes in several parameters. A change in one variable alone may have no real-world counterpart, and as such is not reasonable.

9.4 The Interpretation of Model Response to Changes

A parameter or structural change typically leads to a change in model behavior. How "much" change in model behavior can we accept without feeling a need to increase the precision in the underlying parameters?

How "much" change does it take before we feel we have acquired new understanding about the present structure or identified a sensitive point? There exists no strictly objective answers to questions like these. What constitutes noteworthy sensibility depends on the purpose for which the model was made and also on the user's individual need for precision.

The Role of the Client

Disagreement between the modeler and the client in interpreting the results of a sensitivity analysis may indicate that the two have different perceptions concerning the purpose of the model. The ultimate judgment of the acceptability of the model's sensitivity should rest with the client. This role of the client underscores the subjective nature of sensitivity evaluation.

9.5 Sensitivity Analysis in the Process of System Dynamics Modeling

Sensitivity analysis is often viewed as a test to be performed after the model is complete. In system dynamics modeling sensitivity analysis is spread throughout the model-building process. In general, every model experiment contains an element of sensitivity analysis. Thus sensitivity analysis is part of the model construction process and will often constitute the basis for review and possible redesign of the model. The following two subsections outline the place of sensitivity analysis in the system dynamics modeling process.

Sensitivity Analysis Contained in Model Experimentation

Several stages in modeling involve sensitivity testing in one form or another. The tuning of a system dynamics model can be seen as a limited sensitivity test. This process, which usually concentrates on parameter values, may give the first indications of a model's sensitivity to parameter changes.

Policy analysis can also be viewed as a limited sensitivity test. A policy test is nothing but a change in model structure and/or a change in parameters—a change that is feasible in the actual system, and to which model behavior may be sensitive. Moreover, an important part of policy analysis is testing the robustness of recommended policies with relation to uncertainties in model parameters and structure.

The testing of a model to increase confidence also involves an element of sensitivity testing, and vice versa.

Sensitivity Analysis as a Means to Generate New Dynamic Hypothesis

Model experimentation may generate new dynamic hypotheses in two ways. The first situation arises when the model exhibits a very fast response, compared to the model time horizon, to some change. Typically the model structure does not explain the rapid response but rather assumes that such a response is plausible through some aggregated representation. A modeler might want to make a new model, with a shorter time horizon, to investigate how realistic the aggregate representation is.

The second situation occurs when testing reveals a response generated by very slow processes within the system. Such a response will usually be visible toward the end of the model time horizon. The alternative behavior mode will often suggest another question and inspire the construction of a new model.

The situations can be exemplified by two system dynamics studies. The first example, shown in figure 9.6, illustrates the results of one of the tests carried out on the "solid waste model" (Meadows and Meadows, 1973, pp 165–211). The behavior of interest is that of the market price of raw materials. In the test, a 50 percent increase in extraction cost is imposed on the system in year 25—as a representation of an assumed 50 percent tax on extraction of raw materials. As shown by the figure, the system compensates quickly for the price hike. Approximately five years after implementation of the tax, the price has fallen nearly to the original level. The quick response is caused (in the model system) by lower demand and higher recycling, both caused by the price increase.

Implicit in the model then is the assumption that the recycling rate is able to respond quickly to a change in price. The assumption is reflected in the table function shown in figure 9.7, and in the lack of delay in reaching the indicated values (seen from the model structure illustrated in figure 9.8). The model does not explain the fast response of the recycling rate. On the contrary, it is simply assumed.

Critics might dispute the quick response in the recycling rate. Rather than accepting the assumption, they may call it a hypotheses. To test this dynamic hypotheses would require another model focusing on the mechanisms determining the extent of recycling and with a shorter time horizon of about fifteen years. If the hypotheses were confirmed in the

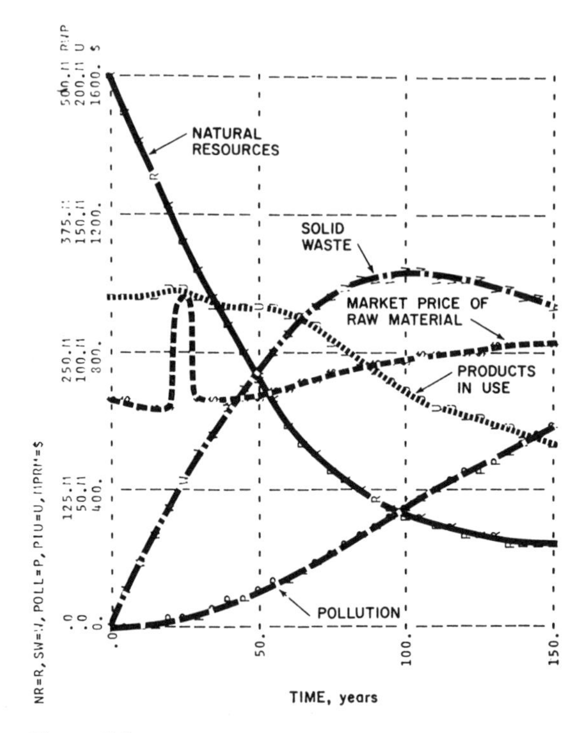

Figure 9.6
The response in market price of raw materials (in dollars) to a 50 percent tax on extraction introduced in the year 25—an example of a fast response compared to the model horizon

new model, the confidence in the original model would increase. Thus an investigation of a new dynamic hypotheses, derived from an original model, may contribute to increased confidence in that model.

The second example is taken from the SOCIOMAD of the interaction between ecosystem (rangeland) and social system (represented by human and animal populations) in the Sahel area (Picardi, 1975). The model purpose is to describe the ecological processes of desertification and the human problem of starvation and herd losses in the Sahel, and to suggest policies to avert the reoccurence of the 1971 to 1974 drought catastrophe.

During comprehensive sensitivity analysis of this model, one experiment revealed a significant shift in model behavior toward the end of the model run (year 2020). In all but this specific test, the most important system variables—soil conditions and animal and human populations—seemed to stabilize at very low levels at the end of the model time horizon. The standard run of the SOCIOMAD model, figure 9.9,

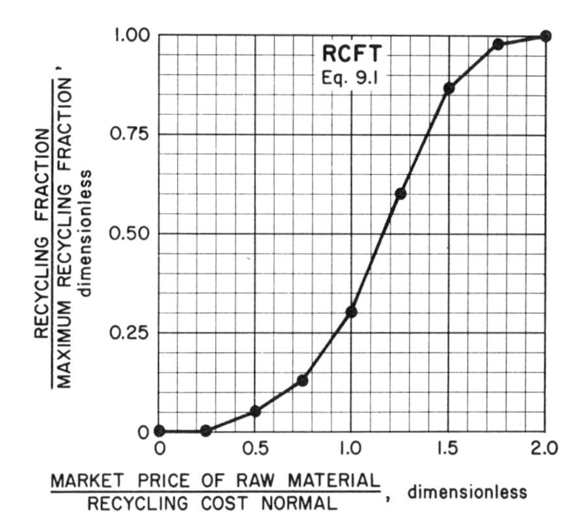

Figure 9.7
The assumed dependence of the recycling fraction on the market price of raw material

illustrates this behavior. In extensive experimentation, it seemed impossible to restore the system to its predrought 1970 levels. A new question emerged: Would it at all be possible to rehabilitate the system after the serious collapse in the early 1970s? Although the question falls beyond the scope of the original model, a deliberate change was made in the assumed future rainfall pattern, aiming at giving the ecosystem a chance to recover. The results of assuming more favorable rainfall patterns are illustrated in figure 9.10. As shown in the figure, the variables are in a growth mode, rather than an equilibrium mode, at the end of this particular simulation. However, the model cannot tell whether such recovery is possible. It only provides a rationale for asking whether such very long-term recovery may take place. To test this hypothesis, we would need a model with a much longer time horizon, explicating the detailed mechanisms that may cause soil to regenerate. Lengthening the time horizon of the original model is no alternative. The SOCIOMAD model has no structure in it capable of answering the new question.

9.6 Final Remarks

The nature of the problems being modeled in system dynamics demands comprehensive sensitivity analysis of the models. In order to judge how

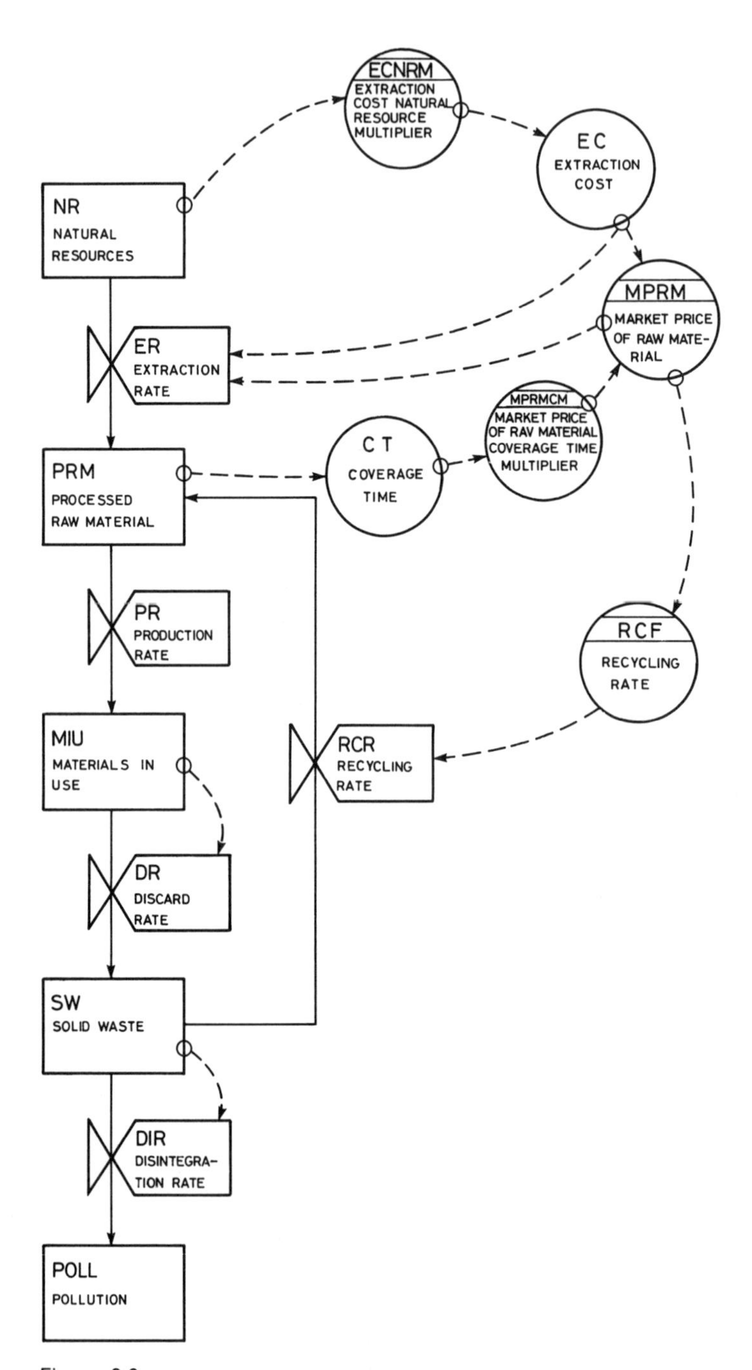

Figure 9.8
Segment of the DYNAMO flow diagram for the solid waste model

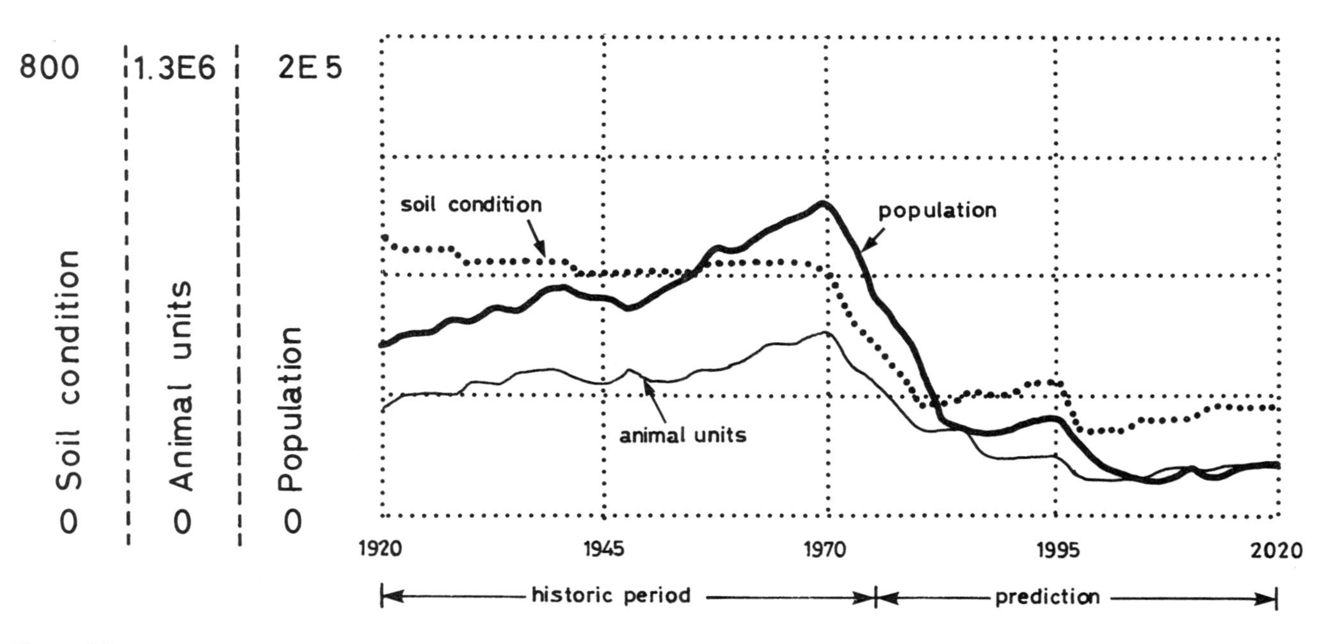

Figure 9.9
The behavior of main variables in SOCIOMAD in the standard run

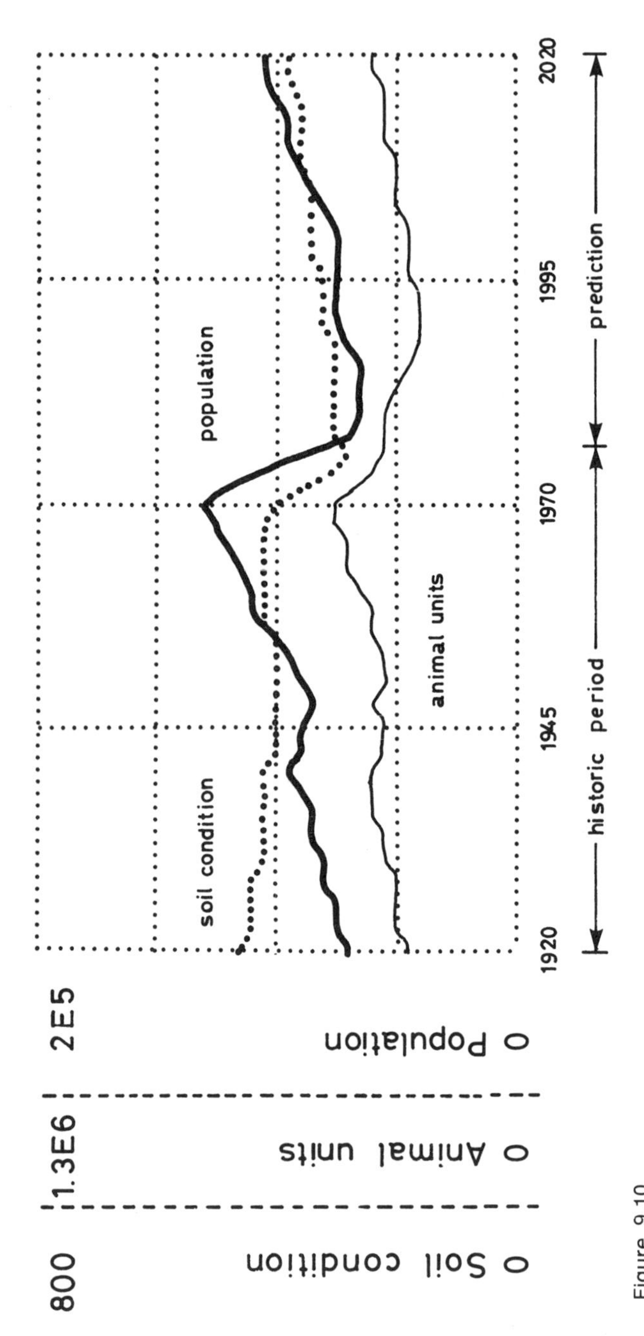

Figure 9.10
The behavior of main variables in SOCIOMAD when more favorable rainfall conditions are assumed

well this demand is met, it is necessary to employ an uncommonly broad perspective on sensitivity analysis. Traditionally, sensitivity testing is performed only on the finished model. In this perspective, we must conclude that system dynamics models normally are subjected to rather moderate sensitivity analyses. Although often stated by critics of system dynamics models, the conclusion is incorrect. In a perspective that truly grasps the role of sensitivity analysis in system dynamics, extensive sensitivity analysis turns out to be the rule. In system dynamics sensitivity testing is spread throughout the modeling process. The extent of sensitivity analysis at each step of the process may be moderate; in sum, however, it is extensive. Sensitivity analysis is in fact a central element of the model construction process.

Note

1. If a model is simplified to the extent that it is difficult to recognize the basic real-world mechanisms at work within the system, its function as a communication tool may be destroyed. Such a simplification is normally not the goal.

References

Derviniotis, E. S., Michelsen, Aa. U., Nielsen, L. E., and Nielsen, O. B. 1973. *System Dynamics, Bind I, Teoretisk og Metodologisk Gjennomgang.* Lyngby, Denmark: Driftsteknish Institut, DTH.

Forrester, Jay W. 1961. *Industrial Dynamics.* Cambridge, Mass.: MIT Press.

Forrester, Jay W. 1969. *Urban Dynamics.* Cambridge, Mass.: MIT Press.

Goodman, Michael R. 1974. *Study Notes in System Dynamics.* Cambridge, Mass.: Wright-Allen Press, Inc.

Graham, A. K. 1976. *"Parameter Formulation and Estimation in System Dynamics Models."* Cambridge, Mass.: System Dynamics Group, MIT, Cambridge, Mass.

Meadows, D. H. and D. L. Meadows, eds. 1973. *Toward Global Equilibrium: Collected Papers.* Cambridge, Mass.: Wright-Allen Press, Inc.

Meadows, Dennis L. et al. 1974. *Dynamics of Growth in a Finite World.* Cambridge, Mass.: Wright-Allen Press, Inc.

Picardi, Anthony C. *"A Systems Analysis of Pastoralism in the West African* Sahel." Ph.D dissertation, MIT, Cambridge, Mass.

Sharp, J. A. 1976. *Sensitivity Analysis. Methods for System Dynamics Models.* System Dynamics Research Group, University of Bradford, Bradford, Yorkshire, England.

Tank-Nielsen, Carsten, and Roger W. A. Brown. 1975. *A Sensitivity Analysis of Picardi's SOCIOMAD.* System Dynamics Group, Dartmouth College, Hanover, N.H.

Alternative Tests for Selecting Model Variables

10

Nathaniel J. Mass
and
Peter M. Senge

10.1 Introduction

One of the most difficult and subtle tasks confronting the mathematical model-builder is selecting appropriate variables and functional relationships. In specifying each equation, the modeler faces two distinct problems. First, he must posit an a priori hypothesis regarding causes of change in the dependent variable of the equation. The a priori hypothesis, which may be based on direct observation, prior theory, or both, identifies the variables believed to be significant determinants of change in the dependent variable. The a priori hypothesis also specifies how these determinants are to be combined. Second, the modeler must have some means of testing whether or not, given the available empirical information, the variables and relationships obtained from a priori reasoning are in fact important. Based on the results of such testing, the modeler may reject certain variables as relatively unimportant, and may thereby begin to refine initial causal hypotheses.

This chapter addresses the second aspect of the problem of variable selection—testing the importance of model variables. The paper contrasts two approaches to model testing—single-equation statistical tests and model behavior tests. Single-equation statistical tests are tests that compare an individual model equation to statistical data. Two such tests are the popular *t*-test of parameter "significance" and the partial correlation coefficient, both of which focus on the impact of one particular "explanatory" variable on a "dependent" variable. Model behavior tests measure the importance of an hypothesized impact of one variable on another for the behavior of a complete system model. The key distinction between the two testing approaches lies in whether or not a closed-loop feedback response is being examined: single-equation statistical tests focus on hypothesized relationships in isolation from the context of feedback relationships in which they are embedded, model behavior tests consider this context.[1]

The major theme of this chapter is that only tests that examine a variable's influence on overall model behavior provide a sound basis for assessing a variable's importance. On the other hand, the single-equation statistical tests employed extensively in modeling practice can be extremely misleading if used as the sole guide for rejecting or accepting a variable in a causal model. Both theoretical and practical arguments are presented in support of the superiority of model behavior tests.

The arguments we present pertain especially to the social sciences, where alternative theories frequently match the statistical evidence equally well. Whereas most social scientists attribute the presence of alternative "equally valid" theories to the paucity of reliable data, the operating philosophy of theory testing may be at least equally at fault.

Criticism of the single-equation statistical testing viewpoint can be traced back to John Maynard Keynes who, in reviewing Jan Tinbergen's *Statistical Testing of Business Cycle Theories: A Method and Its Application to Investment Activity,* argued,

> The method [multiple-correlation analysis] is one neither of discovery nor of criticism. It is a means for giving quantitative precision to what, in qualitative terms, we know already as the result of a complete theoretical analysis. . . . How far are these curves and equations means to be no more than a piece of historical curve-fitting and description, and how far do they make inductive claims with reference to the future as well as the past? If the method [multiple-correlation analysis] cannot prove or disprove a theory, and if it cannot give a quantitative guide to the future, is it worthwhile?[2]

More recently, many social scientists have once again begun to question the power of well-established statistical tests (see, for example, Morrison and Henkel, 1970). Therefore this chapter can be seen as part of a methodological debate which has persisted for over thirty years. Within this context, the chapter is, however, unique in that it prescribes a concrete and broadly applicable alternative to single-equation statistical testing.

10.2 Statistical Tests for the Selection of Model Variables

This section discusses the type of single-equation statistical tests employed frequently in modeling practice as a guide in selecting model variables. The section shows that such tests actually measure the degree to which available data permit accurate estimation of model parameters, and thus should be viewed more as tests of data usefulness than as tests of model specification. A case study in a subsequent section illustrates how, if used as a basis for selecting model variables, the statistical tests can seriously mislead the model-builder. The discussion focuses on two particular statistical measures—the *t* statistic and the partial correlation coefficient.

The *t*-Test[3]

The *t*-test provides the modeler with a measure of the confidence he can place in an estimated parameter value. The test follows from recognizing that an estimated parameter value is one possible value of a random statistical parameter "estimator."[4] The estimator is a random variable because the equation being estimated is assumed to have a random component. Because it is a random variable, the estimator has a mean and a variance. If the estimator has a large variance, little can be said with confidence about its accuracy. That is, even if the mean of the estimator equals the true value of the parameter being estimated (that is, the estimator is "unbiased"), a large variance means that the probability that the parameter estimate is close to the true parameter value is low. The *t*-statistic provides a measure, based upon a formal statistical hypothesis test, to determine whether or not the variance of the parameter estimator is "too large."

In a linear regression, the *t*-statistic is computed as the ratio of the computed parameter estimate $\hat{\beta}$ to the estimated standard deviation of the parameter estimate $\hat{\sigma}_{\hat{\beta}}$:

$$t\text{-stat} = \frac{\hat{\beta}}{\hat{\sigma}_{\hat{\beta}}}. \tag{10.1}$$

As the estimated standard deviation $\hat{\sigma}_{\hat{\beta}}$ increases relative to the parameter estimate $\hat{\beta}$, the *t*-statistic diminishes. Given a certain set of assumptions the *t*-statistic becomes a tool of statistical inference.[5] For example, a *t*-statistic whose absolute value is greater than 2.3 permits the modeler to *infer* with a probability above 0.99 that the true parameter is nonzero.

As a measure of confidence in a paremeter estimate, the *t*-statistic guides the modeler toward judicious use of the available sample of data. Passing the test tells the modeler that an estimate is fairly "tight"—that is, the estimated standard deviation of the estimator is small relative to the estimated parameter value. Failing the *t*-test tells the modeler that the estimated standard deviation of the estimate is large relative to the estimated parameter value. However, neither outcome tells the modeler whether the underlying parameter β or the associated variable is in any sense "important" for the model being estimated. As will be illustrated, the statistical significance or "tightness" of an individual parameter estimate and the importance of the associated variable may differ markedly.[6]

Therefore the t-test is best interpreted as a test of a particular type of data usefulness—not as a test of model specification. Failure to pass the t-test means that the available data do not permit accurate estimation of the parameter β. Conversely, passing the t-test means that the data are useful for the purpose of estimating β—that is, β can be estimated with suitable precision. Although such tests of data usefulness may play an important role in the overall validation process, they must be sharply distinguished from tests of model specification.

The Partial Correlation Coefficient

The partial correlation coefficient provides the modeler with a measure of the incremental contribution of a single right-hand-side ("explanatory") variable in accounting for variation in a dependent variable. Denote the explanatory variable in question x_h. If we compute the fit of the estimated equation twice, once with the variable x_h included in the equation and once with x_h excluded, the partial correlation coefficient, r_h, could be determined from the two computations of the coefficient of multiple determination, R^2 (R_h^2 corresponds to the case when x_h is omitted from the equation):

$$r_h^2 = \frac{R_{\text{total}}^2 - R_h^2}{1 - R_h^2}. \qquad (10.2)$$

An econometrician would say that the partial correlation coefficient measures the "amount of variance explained" by the variable x_h. Although intuitively appealing as a test of the importance of the variable, the partial correlation coefficient actually yields no information not already provided by the t-statistic, as can be seen by the following relationship between the two measures (t_h denotes the t-statistic for the coefficient associated with the variable x_h):[7]

$$r_h = \frac{t_h}{\sqrt{t_h^2 + n - k}}, \qquad (10.3)$$

where r_h = partial correlation coefficient for x_h
t_h = t-statistic for x_h
n = periods of data available
k = number of coefficients to be estimated in the equation.

Equation (10.3) implies that, as the t-statistic approaches zero, the partial correlation coefficient likewise approaches zero. Moreover, the partial correlation coefficient decreases whenever the absolute value of the t-statistic decreases, provided there are more data points than parameters to be estimated ($n - k > 0$).[8]

Because the partial correlation coefficient is so closely coupled to the t-statistic, it may indicate that a variable contributes little in "explaining movements" in a particular dependent variable when, in fact, the relationship between the two variables is simply difficult to measure given available data. Therefore the partial correlation coefficient is clearly not a reliable guide to model specification. To obtain a more reliable measure of the contribution of variable x_h in explaining observed movements in a dependent variable, it is necessary to analyze the behavior of the system of feedback relationships within which the relationship in question is embedded.

10.3 Model-Behavior Tests for the Selection of Model Variables

Given that single-equation statistical measures, such as the t-statistic and the partial correlation coefficient, do not provide reliable measures of the relative importance of different variables, what alternative techniques might provide insight into this critical problem? Section 10.3 attempts to provide some guidelines and direction for addressing the importance of the hypothesized impact of one variable on another. In particular, the discussion focuses on an approach to testing which assess a variable's influence on model behavior.

The model-behavior testing approach entails three principle steps. Suppose the modeler seeks to determine whether or not variable X is an important determinant of observed oscillations in variable Y. First, a model must be constructed which contains enough endogenous structure to portray how changes in one variable, say X, affect the present and future values of both X and Y. The model should generate a pattern of oscillatory behavior in Y similar to that observed in real life.[9] Moreover, model behavior should arise primarily from the model's internal structure, not from exogenous inputs driving the model (Forrester, 1961, ch. 12).

Borrowing from an example developed in section 10.4, suppose the modeler wants to analyze the impact of delivery delay on sales in a firm which has experienced unstable sales growth. An adequate model to address this question should include the hypothesized direct effect of delivery delay on sales effectiveness and, consequently, sales rate (increasing delivery delay reduces sales effectiveness while decreasing delivery delay increases sales effectiveness); but the model should also include the influence of sales (that is, orders) on order backlogs and delivery delay, as well as the influence of sales rate on revenues, capacity expansion, and marketing effort. When simulated, the model should generate the pattern of unstable sales growth observed in the firm. Construction and analysis of a model containing the feedback interactions between production capacity, delivery rate, delivery delay, marketing effort, and sales rate provides the necessary foundation for analyzing the importance of the hypothesized impact of delivery delay on sales.

The second step in whole-model testing involves simulating the model both with and without the direct influence of X on Y. How does the behavior of the variable Y change as a result of deleting the direct link between X and Y? In the delivery delay example, assessing the impact of delivery delay on sales behavior would involve omitting the hypothesized direct link between delivery delay and sales, and then seeing whether model behavior is altered significantly as a consequence.

Finally, the third step in whole-model testing involves analyzing the causes of the behavior observed in the second step. If the behavior of Y is relatively unaltered, what other variables appear to dominate the behavior? If Y's behavior is altered significantly, what direct and indirect links between X and Y account for the change in behavior?

Before proceeding to give specific examples of the model-behavior testing process, a short discussion of alternative criteria for model-behavior testing is appropriate. At least three criteria are possible:

1. Does omission (inclusion) of the factor lead to a change in the predicted numerical values of the system?
2. Does omission (inclusion) of the factor lead to a change in the behavior mode of the system? For example, does it damp out or induce fluctuations in the system?
3. Does omission (inclusion) of the factor lead to rejection of policies that were formerly found to have had a favorable impact or to reordering of preferences among alternative policies?

In general, the results of an evaluation of the importance of an hypothesized impact of one variable on another will depend on which of the three criteria are used. For example, suppose the model designed to explore the causes of sales fluctuations in a particular firm initially exhibits the sales behavior shown by the curve labeled *A* in figure 10.1. Suppose now that omission (or inclusion) of the direct link between delivery delay and sales alters model behavior to that described by curve *B* in figure 10.1. Curve *B* differs from *A* in the exact numerical values for sales over time, but both curves clearly exhibit approximately the same general growth trend and the same magnitude of fluctuations. The difference between outcomes *A* and *B* would then be judged important by the first criterion for model-behavior testing, but unimportant by the second criterion. To deal with the third criterion, suppose that a number of policies were tested on both the models underlying curves *A* and *B* and it was found that the policies that reduce fluctuations in one model also reduce fluctuations in the other, and conversely. In this situation, the difference between the two models would be judged insignificant or unimportant by the third criterion.[10] This example, although highly simplified, illustrates some of the considerations involved in assessing the importance of a given model relationship. The purpose of a particular study will, in general, determine which of the three criteria is appropriate for the evaluation process.

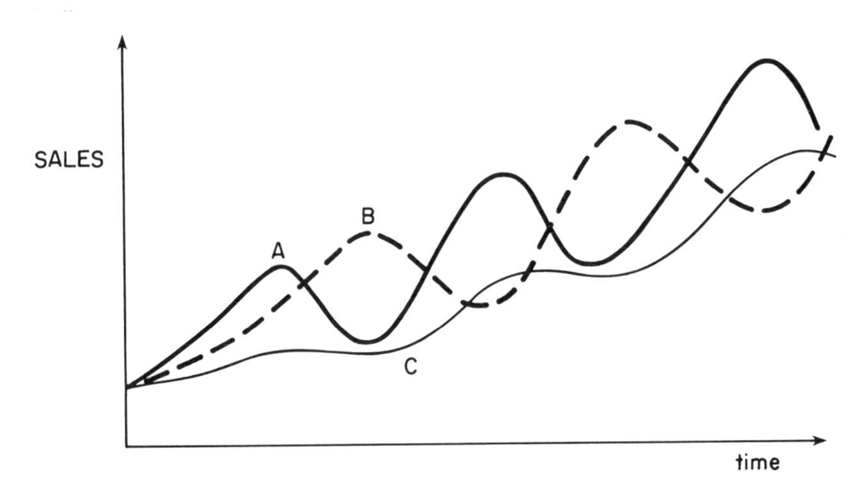

Figure 10.1
Alternative patterns of sales behavior

10.4 A Case Study in Selecting Model Variables

The following example demonstrates the operation and consequences of the two previously described types of testing. The example shows, first, that model-behavior testing yields information about a variable's influence on behavior that cannot be ascertained by statistical testing and, second, that the information produced by model-behavior testing can aid in discriminating which variables are important in generating particular patterns of behavior.

The Model

The following example utilizes a fairly simple feedback model built by Forrester, 1968, to explain how a rapidly growing firm can experience instability in sales even in the presence of potentially limitless demand. Figure 10.2 provides an overview of the structure of the "market-growth" model. The model assumes that the firm expands or contracts its sales force depending upon the difference between the number of

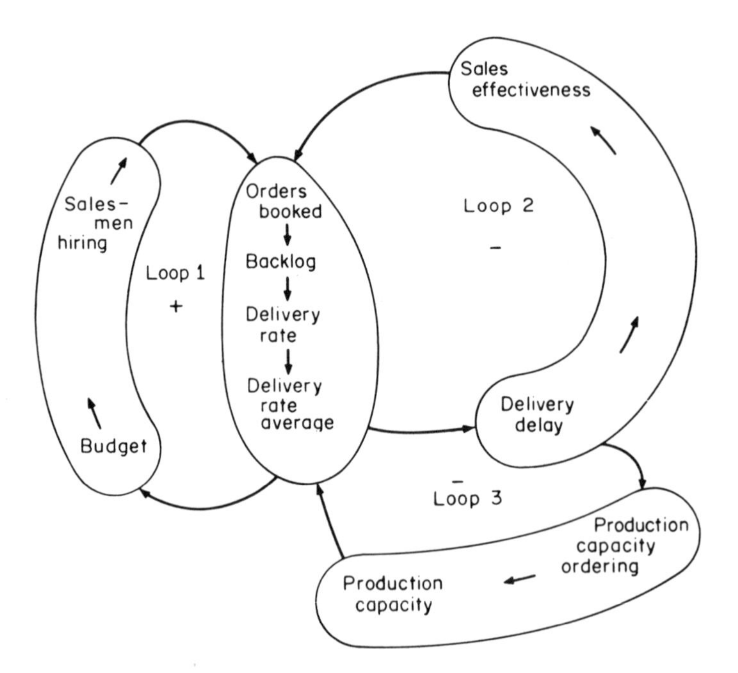

Figure 10.2
Major feedback loops in Forrester's market-growth model

salesmen that can be supported by the marketing budget and the existing sales force. The firm orders additional production capacity when delivery delay becomes longer than desired (the firm's desired delivery delay is taken as two months) and reduces capacity when delivery delay falls below the desired value. Delivery delay also influences sales (orders booked) through sales effectiveness. Assuming a highly competitive market, the model hypothesizes that customers reduce orders whenever delivery delay rises, and vice versa.

Given such a model of interactions within a firm, how could we test the hypothesized impact of delivery delay on sales? Assume that the real-life firm has experienced rapid growth in sales and production capacity punctuated by intermittent sharp declines in sales. Assume also that ample data are available for order backlog, delivery delay as recognized by the market, salesmen, and production capacity.

Statistical Tests of the Hypothesized Impact of Delivery Delay on Sales

In order to test statistically the hypothesized impact of delivery delay on sales, we must write out the appropriate equations, identifying unknown parameters to be estimated. The equations (10.4) through (10.8) taken from the original Forrester model, describe the determinants of change in order backlog, orders booked (sales), and delivery rate. Random terms are included in the equations for orders booked and delivery rate:

$$\Delta BL = OB_t - DR_t \tag{10.4}$$

$$OB_t = S_t \cdot SE_t + \epsilon_{1_t} \tag{10.5}$$

$$SE_t = g_1(DDRM_t) \tag{10.6}$$

$$DR_t = PC_t \cdot PCF_t + \epsilon_{2_t} \tag{10.7}$$

$$PCF_t = g_2(BL/PC)_t \tag{10.8}$$

$$\Delta BL = BL_{t+1} - BL_t,$$

where

BL = order backlog (units)

DR = delivery rate (units shipped per month)

SE = sales effectiveness (units sold per salesman per month)

PC = production capacity (maximum possible units shipped per month)

$DDRM$ = delivery delay recognized by market (months)

PCF = production capacity fraction (dimensionless)

ϵ_1, ϵ_2 = random error processes

$g_1(°)$ = nonlinear function

$g_2(°)$ = nonlinear function.

In order to estimate equation (10.4), the order backlog equation, the nonlinear functions g_1 $(DDRM)$ and g_2 (BL/PC) must be parameterized. In the following experiments, the function $g_1(°)$, which incorporates the impact of delivery delay on sales, is approximated as linear and $g_2(°)$ is specified as a third-order polynomial ($g_2(°)$ is definitionally constrained to be zero when (BL/PC) equals zero).[11] After rearranging, delivery delay recognized by market $DDRM$, the variable whose impact is being tested, enters in the second coterm multiplied by the parameter $K2$:

$$\begin{aligned}
\Delta BL &= S_t \cdot (K1 + K2 \cdot DDRM_t) + PC_t \cdot (K3 \cdot (BL/PC)_t \\
&\quad + K4 \cdot (BL/PC)_t^2 + K5 \cdot (BL/PC)_t^3) + (\epsilon_{1t} + \epsilon_{2t}) \\
&= K1 \cdot S_t + K2 \cdot S_t \cdot DDRM_t + K3 \cdot BL_t \\
&\quad + K4 \cdot (BL^2/PC)_t + K5 \cdot (BL^3/PC_t^2) + \epsilon_t \qquad (10.9) \\
\Delta BL &= BL_{t+1} - BL_t,
\end{aligned}$$

ϵ_t is assumed to be zero-mean, normally distributed, stationary, and white. $K1, \ldots, K5$ are unknown parameters. According to prior reasoning, increases in delivery delay should suppress sales, hence the sign of $K2$ should be negative.

To test statistically the hypothesis that delivery delay influences sales, we could examine the t-statistic for the parameter $K2$ or the partial correlation coefficient for the coterm $S_t \cdot DDRM_t$ in equation (10.9). In order to examine the performance of the statistical tests, we conduct the following simple experiment. Simulating the entire market-growth model, we generate "synthetic" data for the variables involved in equation (10.9) and use that data for estimating the equation. That is, we operate on model-generated ("synthetic") data just as the econometrician would operate on real data. The synthetic data experiment provides a useful framework for examining the performance of statistical tests because, within the experimental framework, we know the structure of the "true" data-generating system perfectly. Such synthetic data experiments are common in the econometrics and statistics literature on evaluating estimation techniques and have been recently used to evaluate estimators for feedback models (Senge, 1977).

First, an experiment is conducted under the highly idealized conditions of perfect measurement of all model variables. As shown in the first estimation in table 10.1, when data measurement is perfect, statistical estimation (using ordinary least squares, OLS) results in a statistically significant estimate for the parameter $K2$ (t-statistic equal to –9.69) and a high partial correlation coefficient (–0.7049). However, when moderate measurement errors are permitted to enter the data, the statistical results

Table 10.1
Statistical tests for order backlog equation (ordinary least-squares estimation)[a]

Coef-ficient	True value	Estimated value	$\hat{\sigma}_{\hat{\beta}}$	t-stat	r_h	
Error-free data						
$K1$	475	457.7	37.05	12.36	0.7851	
$K2$	-61.5	-54.62	5.638	-9.69	-0.7049	
$K3$	-0.6178	$-.6484$	0.06398	-10.13	-0.7207	$R^2 = 0.9934$
$K4$	0.1324	0.136	0.02018	6.74	0.5689	
$K5$	-0.00975	-0.00975	0.00299	-3.26	-0.3171	
Error-corrupted data						
$K1$	475	408.2	89.48	4.562	0.4239	
$K2$	-61.5	-26.85	21.51	-1.248	-0.1270	
$K3$	-0.6178	$-.5563$	0.1530	-3.637	-0.3496	$R^2 = 0.8486$
$K4$	0.1324	0.07719	0.05362	1.440	0.1461	
$K5$	-0.00975	-0.00411	0.00606	$-.6792$	-0.0695	

[a] $BL = BL_{-1} + K1 \cdot S_{-1} + K2 \cdot S_{-1} \cdot DDRM_{-1} + K3 \cdot BL_{-1} + K4 \cdot (BL_{-1}^2 / PC_{-1}) + K5 \cdot (BL_{-1}^3 / PC_{-1}^2)$.

are adversely affected, as shown in the second estimation in table 10.1.[12] The t-statistic (−1.248) indicates that the estimated delivery delay impact is statistically insignificant. Moreover, the coterm $S_t \cdot DDRM_t$ exhibits low partial correlation relative to the partial correlation for the coterms $K1$, $K3$, and $K4$. Overall, the statistical tests based on the error-corrupted data give no evidence to support the hypothesized influence of delivery delay on sales, even though, by the very design of the computer experiment, a *direct impact of delivery delay on sales* is present in the data-generating model. Therefore the statistical tests are not reliable tests of the specification of the order backlog equation. What the tests do indicate is that, when the quality of available data becomes poor, the hypothesized impact of delivery delay on sales becomes difficult to estimate. In this sense, the statistical tests measure the usefulness of the available data rather than the specification of the market-growth model.

Model-Behavior Test of the Hypothesized Impact of Delivery Delay on Sales

The preceding section showed that, even with perfect knowledge of the determinants of sales, statistically insignificant estimates of the influence of delivery delay on sales may be obtained when data contain moderate measurement errors. An econometrician viewing this result might conclude that delivery delay is a relatively unimportant influence on sales. However, using model-behavior testing, delivery delay actually exerts a pronounced influence on sales behavior in the market-growth model.

Figure 10.3 shows the basic behavior of the market-growth model when the direct link between delivery delay and sales, shown in figure 10.2, is present. (To simplify the discussion, the simulations in this section do not include random components.) Figure 10.3 shows a behavior pattern of growth in sales (orders booked), which is interrupted by periods of decline. Such growth instability is caused by the simultaneous effect of delivery delay on sales and capacity expansion. At the outset of the simulation, the firm's delivery delay is low. Sales effectiveness is consequently high, thereby allowing orders booked to increase rapidly as the sales force increases. The growth of sales increases the firm's order backlog. Delivery delay rises during this period because order backlog is growing more rapidly than production capacity. Eventually, delivery delay increases to the point where high delivery delay

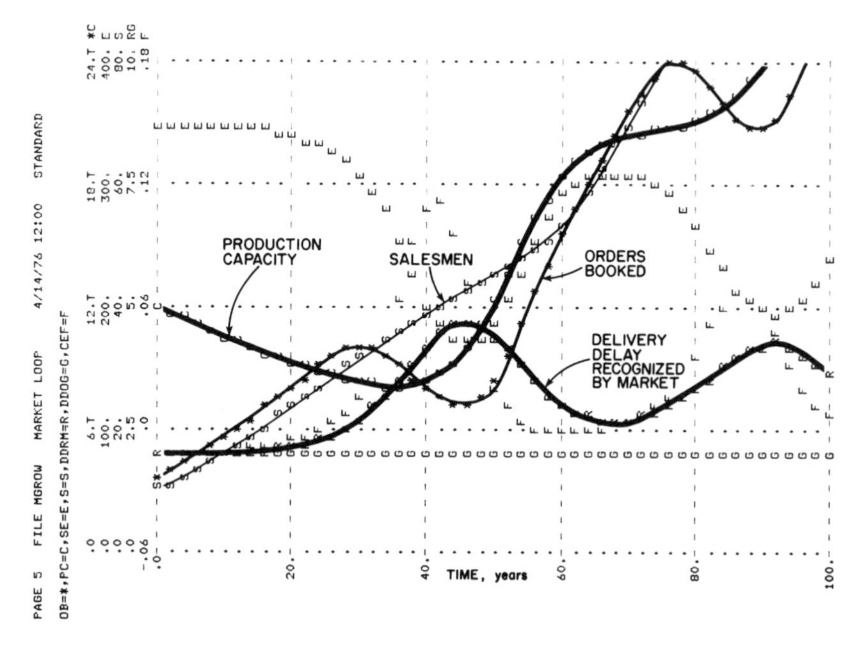

Figure 10.3
Standard run of market-growth model

begins to suppress orders. As the firm perceives the long delivery delay, it begins to order new production capacity. However, due to delays in perceiving delivery delay and in acquiring production capacity, delivery delay as perceived by the market continues to rise for a while, thereby lowering sales effectiveness and depressing orders booked. As capacity eventually begins to expand, delivery delay starts to decline. The decline in orders booked and order backlogs further lowers delivery delay. Once delivery delay falls, growth in orders booked resumes and the rate of capacity expansion declines. Figure 10.3 therefore shows alternating periods of growth and decline in sales and production capacity. In the following discussion, the pattern of system behavior shown in figure 10.3 will be assumed to be the pattern of real-life behavior which the modeler seeks to reproduce and understand.

In order to test the contribution of delivery delay to sales behavior, the direct impact of delivery delay on salesman effectiveness can be eliminated from the market-growth model. A simulation of the resulting model is shown in figure 10.4.[13] By contrast to the previous simulation, figure 10.4 shows continued growth in orders booked and continued increase in production capacity after month 30, although both orders and

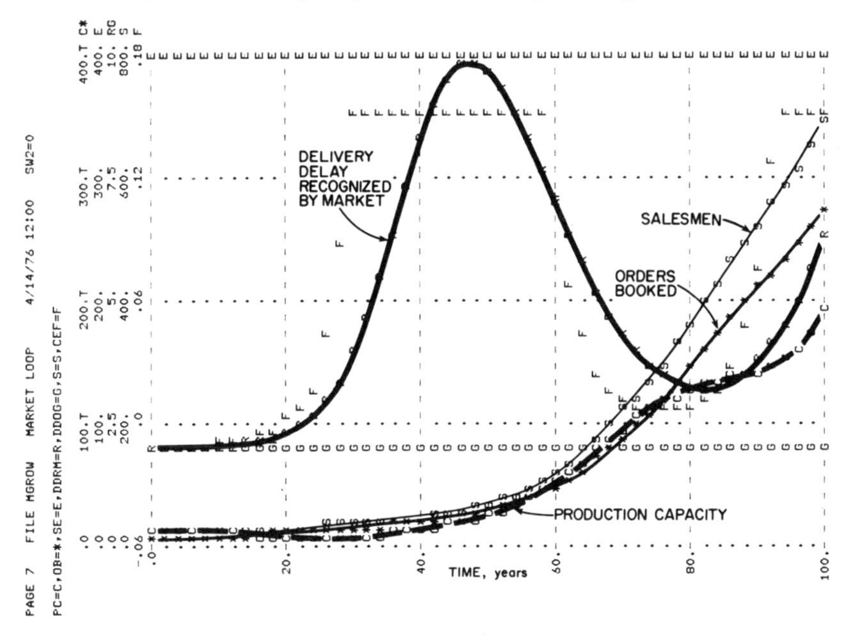

Figure 10.4
Market-growth model with no direct influence of delivery delay on orders booked (sales)

capacity expand at a fluctuating rate of growth. Therefore, with the direct link between delivery delay and sales eliminated, the market-growth model no longer exhibits the recurring fall-offs in sales characteristic of the original model. With this link omitted, a high delivery delay no longer exerts a depressive effect on sales. Instead, sales are influenced only by the level of salesmen and by a constant sales effectiveness.

Analysis of the basic feedback structure of the market-growth model, depicted in figure 10.2, clearly shows why omission of the influence of delivery delay on sales shifts model behavior from the pattern seen in figure 10.3 to the outcome in figure 10.4. Loop 1 in figure 10.2 is a positive feedback loop. Within the loop, an increase in delivery rate raises revenues; high revenues in-turn will increase the marketing budget, leading to increased salesmen, increased orders, and, thereby, still higher sales. Loop 1 therefore tends to promote continued exponential growth in sales.

Loops 2 and 3 are negative feedback loops controlling delivery delay. In loop 2, an increase in orders raises order backlog, causing delivery delay to increase. In turn a higher delivery delay lowers salesman effectiveness, thereby lowering sales. In loop 3, an increase in delivery delay encourages capacity expansion, causing an increase in delivery rate and a decline in delivery delay.

When delivery delay is omitted as a direct influence on sales rate, loop 2 no longer constrains sales growth. Sales behavior instead is controlled by the positive feedback loop, loop 1. Inactivation of loop 2 therefore accounts for the nearly steady exponential growth in sales observed in figure 10.4. In figure 10.4, production capacity grows continually after month 30 (due to steadily rising sales), but at a fluctuating rate of increase. These fluctuations in production capacity expansion are generated in loop 3, and are coupled to the fluctuations in delivery delay seen in figure 10.4.

Comparison of figures 10.3 and 10.4 shows the critical impact of delivery delay on sales behavior. When delivery delay is no longer assumed to influence sales, the behavior of the market-growth model is altered markedly, and the model no longer exhibits alternating periods of growth and decline in sales. In other words, the overall behavior mode of the system is significantly changed when the delivery delay effect is omitted. Model-behavior testing therefore shows that the hypothesized impact of delivery delay on sales exerts an important influence on model behavior. This result is especially striking in light of the outcome of the statistical tests in table 10.1, which showed that delivery delay was not a statistically significant determinant of sales. The outcomes of the alternative tests highlight the point that only the model-behavior test can be viewed as a true measure of the importance of the hypothesized impact of delivery delay on sales.

10.5 Applications of Model-Behavior Testing to Alternative Theories of Social Behavior

In a recent application, Mass (1975, 1976) used whole-model testing to assess the generic causes of short-term and medium-term cycles in a national economy. In the study, he develops a general model of a producing unit within the economy, and then modifies the production model to incorporate various hypothesized causes of economic instability.

One particularly significant outcome of the study concerns the relative importance of labor adjustments and fixed capital investment in generating short-term business cycles. As noted by Burns (1969) the predominant number of business-cycle theories, including the theories of Paul Samuelson, John Hicks, Nicholas Kaldor, and James Duesenberry, emphasize fluctuations in the flow of fixed capital investment as a cause of overall fluctuations in income and output.[14] Such theories have been widely influential from a theoretical standpoint, and have stimulated much subsequent business-cycle research. They have gained further support in the form of statistical studies showing relatively large swings in investment over a typical business cycle (see, for example Evans, 1969). Moreover, widespread acceptance of the theories has led to economic stabilization policies that emphasize regulating investment opportunities.

Mass employs model-behavior testing to evaluate the widely held hypothesis that fluctuations in capital investment are essential to the business cycle. In his first test, Mass holds fixed capital stock constant while labor is allowed to vary. The resulting simulation exhibits a four-year fluctuation resembling the short-term business cycle in terms of amplitude, phase relationships between variables, and other characteristics. This simulation indicates that a short-term business cycle can be generated independent of fluctuations in fixed capital investment.

In a second test, labor (employment) is held constant while capital stock is permitted to vary. The resulting simulation exhibits a fluctuation of around eighteen-year periodicity resembling the so-called Kuznets cycle (see Abramovitz, 1961). The outcome suggests that variations in fixed capital investment alone, without variations in employment, cannot account for the occurrence of short-term business cycles in the economy, and probably underlie much longer-term cycles.

Similar model-behavior testing might be employed to address some of the major controversies in economic theory and policy. For example, most attempts to date to evaluate the monetarist theories regarding economic cycles and stabilization policy have utilized single-equation statistical tests (for example, Andersen and Jordan, 1968). Reflecting considerations similar to those described above, Blinder and Solow (1974) have argued that the single-equation approach has the theoretical weakness that it ignores the feedback from changes in income to changes in fiscal and monetary policy. A more fruitful approach for evaluating the monetarist theories might be to incorporate the monetarist assumptions into a broad national model to evaluate their implications, significance, and interaction. Such a model might draw, for example, on Friedman's 1968 presidential address to the American Economic Association (Friedman, 1968), in which he presents a descriptive summary of some of the main relationships linking money supply, interest rates, GNP, prices and price expectations, and capital investment according to the monetarist view. Such a model should help to define and assess these assumptions in a more comprehensive framework than has been available to date.

10.6 Conclusions

This chapter has examined two approaches for determining whether or not an hypothesized impact of one variable upon another should be included in a model. The first method analyzed was the single-equation statistical testing approach. The second approach entailed the analysis of model behavior. The major finding of the chapter is that, of the two basic approaches, *only* behavior tests provide a valid basis for selecting model variables. Only by analysis of model behavior can the modeler ascertain the importance of a particular variable. He can do so by omitting any influence of the variable from the model, or by constraining the variable's movement, and examining the consequent shift in model behavior. Model-behavior testing can be used to isolate the influence of an individual variable on a particular historical behavior pattern, on a possible mode of future behavior, or on model response to alternative policies.

By contrast, the proper role of single-equation statistical tests is much narrower. For example, the widely used t-test and partial correlation coefficient provide information only on the precision with which a given parameter can be estimated, not on the importance of the parameter or the associated variables. An example in section 10.4 showed how an hypothesized impact of delivery delay on sales in a firm could fail the statistical tests, even though the delivery delay impact was crucial·for the behavior of the particular model in question. Given the narrower focus of the statistical tests, they should be viewed more as tests of data usefulness than as tests of model specification. That is, failure to pass the statistical tests should not lead the modeler to reject the hypothesized relationship in question, but rather to recognize that the hypothesis is difficult to measure from the available data. Conversely, passing such tests does not mean that the hypothesis in question is in any sense "important," only that it is measurable. If indicators of measurability are used as guides to model specification, they can lead to rejection of relationships that are extremely important for system behavior.

Future research should endeavor to delineate further the possible uses and misuses of statistical and model-behavior testing. A number of theoretical debates in the social sciences might be resolved or at least greatly clarified through application of the behavior-testing approach outlined here. The National Model currently being constructed in the System Dynamics Group at MIT (see Forrester, Mass, and Ryan, 1976) should provide a powerful framework for analyzing many of the more persistent controversies in economics, such as the role of capital investment and monetary policy in economic stabilization, and the monetarist-fiscalist debates. Future research should also focus on possibilities for integrating the two testing approaches. Are there, for example, circumstances under which the results of single-equation tests provide a useful input to behavior testing? Such questions can probably only be answered within the context of a well-defined validation problem, such as the comparison of alternative investment formulations currently being conducted as part of the above-mentioned national modeling project at MIT (see Senge, 1976).

Continued research aimed at developing the model-behavior-testing approach and integrating it more fully with more established testing approaches may contribute to a basic reorientation of model building and theory testing in the social sciences. Although the limitation of statistical testing have been well understood for some time, modeling practice continues to be dominated by the statistical testing perspective. Model-builders continue to reject hypotheses on the basis of low statistical significance (see Goldfeld, 1973, and Schultz, 1969, for example), and to attribute inconclusive statistical studies to poor data bases, as if more complete data would enable the statistical testing methodology to discriminate successfully among alternative hypotheses. This continuing reliance on statistical tests is closely linked to the heritage of single-equation models common to all the social sciences. To the extent that researchers still construct single-equation models, model-behavior testing is not possible. In fields such as econometric modeling, the emergence of system models has preceded the adoption of a commensurate approach to model testing; consequently, econometricians tend to build multiple-equation models within an essentially single-equation philosophy of model testing. This is unfortunate, for such practice overlooks one of the greatest strengths inherent in the systems approach—to test the effect of alternative hypotheses on a complete system, just as is done in the experimental sciences. Hopefully, increased understanding of basic issues such as those raised in this chapter will begin to foster a new philosophy and approach for theory testing in the social sciences.

Notes

1. Most statistical tests employed in econometric practice belong to the single-equation class, regardless of whether the statistical test is based on a recursive, simultaneous, or even multiple equation (for example, Zellner, 1963), estimator. On the other hand, statistical tests using an estimator based on the Kalman filter (see Kalman, 1960, Schweppe, 1973, or Peterson, 1975) belong to the class of model behavior tests. To see the distinction, consider that the former set of estimators do not involve simulating (or, in the case of an extremely simple system, analytically solving for) model behavior in computing a likelihood function, while a Kalman filter-based estimator does.

2. Keynes, 1939, pp. 567, 569.

3. The following exposition of the *t*-test does not attempt to explain fully the mechanics of the test. For such explanation, the reader should refer to an introductory econometrics text (for example, see Theil, 1971).

4. The term "estimator" refers to the estimation technique as an operator which converts a set of data into a set of parameter estimates. The distinction between the estimator and the estimate is analogous to the distinction between a random variable and one particular value of the random variable.

5. The assumptions underlying the *t*-test include perfect specification of the model being estimated (including zero mean and normally distributed noise inputs in each equation) and perfect measurement of all variables. Senge, 1977, discusses the realism of these assumptions and how typical violations of the assumptions affect the accuracy of least squares parameter estimates.

6. For example, multicollinearity and measurement error are two common causes of statistical insignificance. Neither multicollinearity, which arises when one variable on the right-hand side of a regression is highly correlated with another right-hand-side variable, or with a linear combination of right-hand-side variables, nor measurement error necessarily imply that the model itself is defective as a causal description of the real system or that individual model variables are "unimportant."

7. Thiel, 1971, p. 174, provides a derivation of equation (10.3).

8. This can be seen by examining the partial derivative of r_h with respect to t_h:

$$\frac{\partial r_h}{\partial t_h} = \frac{(n-k)}{(t_h^2 + n - k)\sqrt{[t_h^2 + n - k)}}.$$

This equation shows that $\partial r_h/\partial t_h$ is greater than zero provided $(n - k) > 0$.

9. Forrester, 1961, chapter 13 discusses particular aspects of oscillatory behavior, such as average periodicity and phase relationships between variables which provide valid measures of the correspondence between model-generated oscillations and observed oscillations.

10. To show still a more complex case, consider curve *C* in figure 10.1. Curve *C* shows alternating periods of growth and leveling off rather than growth and decline as in curves *A* and *B*. Curve *C* might be judged to be significantly different from, say, curve *A* by both the first and second criteria. The outcomes might not be significantly different from the standpoint of the third criterion, however, if, for example, the same policies that reduce fluctuations in *A* also contract the leveling-off periods in *C*, and conversely.

11. Approximating $g_1(\cdot)$ as linear and $g_2(\cdot)$ as a third-order polynomial allows the modeler to draw the maximum statistically significant information from the experimental data. The variation in delivery delay recognized by the market *DDRM* does not carry that variable into significantly nonlinear regions of the relationship $g_1(\cdot)$. The opposite holds for the backlog to production capacity ratio *(BL/PC)*, which ranges well into the nonlinear regions of $g_2(\cdot)$.

12. Random errors with standard deviations equal to 10 percent of the current value of the error-free data are present in table 10.1. Errors of 5 to 10 percent are typical in economic data according to Morgenstern, 1963.

13. Note that some vertical plot scales are different in figures 10.3 and 10.4 due to the expanded range of variation in salesmen, orders booked, and production capacity in figure 10.4.

References

Abramovitz, Moses. 1961. "The Nature and Significance of Kuznets Cycles." *Economic Development and Cultural Change*, vol. 9 (April).

Andersen, L. C., and J. L. Jordan. 1968. "Monetary and Fiscal Actions: A Test of their Relative Importance in Economic Stabilization." Federal Reserve Bank of St. Louis *Review,* vol. 52 (November), pp. 11–24.

Blinder, A. S., and R. M. Solow. 1974. "Analytical Foundations of Fiscal Policy." In *The Economics of Public Finance.* Washington, D.C.: The Brookings Institute, pp. 67–70.

Burns, Arthur F. 1969. *The Business Cycle in a Changing World.* New York: National Bureau of Economic Research.

Deusenberry, James S. 1969. *Business Cycles and Economic Growth.* New York: McGraw-Hill.

Evans, Michael K. 1969. *Macroeconomic Activity.* New York: Harper and Row.

Forrester, Jay W. 1961. *Industrial Dynamics.* Cambridge, Mass.: MIT Press.

Forrester, Jay W. 1968. "Market Growth as Influenced by Capital Investment." *Industrial Management Review,* vol. 9 (Winter).

Forrester, Jay W., Nathaniel J. Mass, and Charles J. Ryan. 1976. "The System Dynamics National Model: Understanding Socio-Economic Behavior and Policy Alternatives." *Technological Forecasting and Social Change,* vol. 9 (July).

Foster, Richard O. 1972. "Education in the City." System Dynamics Group working paper D-2144. Alfred P. Sloan School of Management, MIT, Cambridge, Mass.

Friedman, Milton. 1968. "The Role of Monetary Policy." *American Economic Review,* vol. 58 (January), pp. 1–17.

Goldfeld, Stephen J. 1973. "The Demand for Money Revisited," *Brookings Papers on Economic Activity,* vol. 3.

Hicks, John R. 1949. "Mr. Harrod's Dynamic Theory." *Economica,* vol. 16 (May), pp. 106–121.

Kaldor, Nicholas. 1940. "A Model of the Trade Cycle." *Economic Journal,* vol. 50 (March), pp. 87–92.

Kalman, Rudolf E. 1960. "A New Approach to Linear Filtering and Prediction Problems." *Journal of Basic Engineering,* series D, vol. 82 (March).

Keynes, J. M. 1939. "Professor Tinbergen Method." A review of J. Tinbergen, *Statistical Testing of Business Cycle Theories: A Method and Its Application to Investment Activity.* Geneva: League of Nations, 1939. *Economic Journal* (September), pp. 567, 569.

Mass, Nathaniel J. 1975. *Economic Cycles: An Analysis of Underlying Causes.* Cambridge, Mass.: MIT Press.

Mass, Nathaniel J. 1976. "Modeling Cycles in the National Economy." *Technology Review,* vol. 78 (March–April).

Morgenstern, O. 1963. *On the Accuracy of Economic Observations.* Princeton, N.J.: Princeton University Press.

Morrison, Denton E., and Ramon E. Henkel, eds. 1970. *The Significance Test Controversy*. Chicago: Aldine Publishing Co.

Peterson, David W. 1975. "Hypothesis, Estimation, and Validation of Dynamic Social Models." Ph.D. dissertation. Department of Electrical Engineering, MIT, Cambridge, Mass.

Roberts, Nancy. 1974. "A Computer System Simulation of Student Performance in the Elementary Classroom." *Simulation and Games,* vol. 5 (September), pp. 265–289.

Samuelson, Paul A. 1939. "Interaction Between the Multiplier Analysis and the Principle of Acceleration." *Review of Economics and Statistics,* vol. 21 (May), pp. 75–78.

Schultz, T. Paul. 1969. "An Economic Model of Family Planning and Fertility." *Journal of Political Economy,* vol. 77 (April).

Schweppe, Fred. 1973. *Uncertain Dynamic Systems.* Englewood Cliffs, N.J.: Prentice Hall.

Senge, Peter M. 1976. "The System Dynamics National Model Investment Formulation: A Comparison to the Neoclassical Model." System Dynamics Group working paper D-2431. Alfred P. Sloan School of Management, MIT, Cambridge, Mass.

Senge, Peter M. 1977. "Statistical Estimation of Feedback Models." *Simulation,* vol. 28 (June).

Theil, H. 1971. *Principles of Econometrics.* New York: John Wiley and Sons.

Zellner, Arnold. 1963. "An Efficient Method of Estimating Seemingly Unrelated Regressions and Tests for Aggregation Bias." *Journal of the American Statistical Association,* vol. 57, pp. 977–992.

Statistical Tools for System Dynamics

11

David W. Peterson

11.1 New Capabilities for System Dynamics

For questions of parameter choice and validity, the system dynamicist has usually relied on "manual" examination of the detailed structure of the model. The realism of both parameter values and model structure is assessed and improved by repeated simulation experiments. If a simulation experiment reveals something surprising or wrong, the modeler asks why, seeks the answer by examining the model structure, and tests the answer by new simulations. This informal procedure of model inspection and simulation is one of the great strengths of the system dynamics methodology. If the modeler proceeds with diligence and thoroughness, the model is greatly improved over "first cut" form, and the modeler gains a deep understanding of the system being modeled.

Numerical data contribute to the process, but usually only when the implications of the data are obvious by inspection. While econometric methods are sometimes employed by system dynamicists (see Hamilton, 1969, and Runge, 1976, for examples), such use is infrequent. The methods often restrict the form and content of a model according to whatever data is available. Furthermore, system dynamicists have sometimes doubted the validity of standard econometric tools. For example, Senge (1974) shows that generalized least squares estimation, *GLS*, may give results that are both wrong and misleading when used in the context of a system dynamics model.

This chapter describes a relatively new set of statistical tools which permit the system dynamicist to make full use of numerical data. The tools relate the model to all relevant data, even if the data is incomplete, noisy, and marred by occasional "bad data" points. The tools work with models in which most of the variables are unmeasured, and for cases of unmeasured exogenous inputs. The tools work correctly under the circumstances examined by Senge and are helpful in answering such questions as,

1. What are the most likely values of unknown parameters, given available data?
2. Which structural formulations are most likely?
3. Is the model consistent with all available data?
4. Which data points are likely to be wrong?
5. What is the most likely state of the system at a given time?
6. To what degree of accuracy can model-computed forecasts be trusted?

The tools are based on the method of full-information maximum likelihood via optimal filtering, *FIMLOF,* as discussed in Schweppe (1973) and Peterson (1975). The mathematics of *FIMLOF* are stated and referenced in this chapter, but the formal proofs will not be repeated here. Instead the dynamic ideas behind the methods are emphasized, and related both to alternate methods and to practices in system dynamics. The tools have also been implemented in a computer program, GPSIE (the general purpose system identifier and evaluator, as described in Peterson, 1974).

Parameter Estimation

The *FIMLOF* method of parameter estimation is best understood as an optimum compromise between two less satisfactory extremes. One extreme might be called "naive simulation," *NS,* and the other extreme is the standard econometric tool, ordinary least squares, *OLS.* To make the presentation clear, this discussion will focus on an extremely simple system; but the method also succeeds with nonlinear, multistate variable systems.

Consider the system

$$X(t) = A*X(t-1) + W(t)$$

$$Z(t) = X(t) + V(t),$$

where X is the state of the system, Z are measurements of X, A is an unknown parameter, $W(t)$ is driving noise ("equation error"), and $V(t)$ is measurement noise ("errors in the variables").

First, take the case where $W(t) = 0$ and $V(t) = 0$, which is equivalent to perfect measurement of a deterministic system. From the data given in figure 11.1, in this noise-free case the parameter A can be estimated by simply taking the ratio between two successive values of $Z(t)$. However,

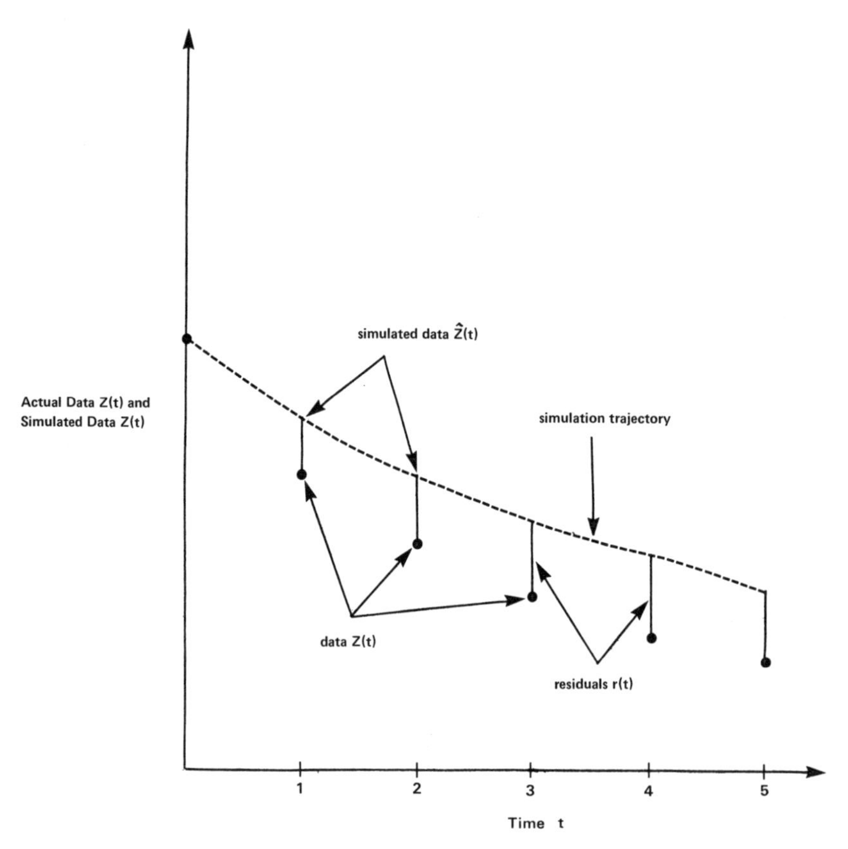

Figure 11.1
Completely deterministic data with *NS* estimates

the simple example can help to illustrate more indirect methods, which succeed not only in the noise-free case but also in more complicated situations. The essence of the more indirect methods, which lead to *FIMLOF,* is to (1) guess a value of A and simulate the system, (2) measure the error between the simulated data $\hat{Z}(t)$ and the actual data $Z(t)$, and (3) repeat the process, making new guesses of A in an orderly fashion, until a value of A is found to minimize the error. The estimate of A is then chosen as the value which minimizes the error between actual and simulated data.

The differences among *NS, OLS,* and *FIMLOF* lie mainly in how the simulation is performed, and in the measure of the error.

Naive Simulation In the naive simulation method, the model is initialized at the first data points and simulated without further reference to the data:

$$\hat{X}(t) = A*\hat{X}(t-1)$$

$$\hat{Z}(t) = \hat{X}(t).$$

In general, the simulated values will not coincide with the data; the differences are called residuals:

$$r(t) = Z(t) - \hat{Z}(t).$$

The *NS* sum of squared residuals, also called the loss function, is denoted as J:

$$J = \sum_{t=1}^{N} r^2(t).$$

In the nonpathological case of figure 11.1, the error will be zero if A is guessed correctly. The modeler may guess close to the right A value more efficiently and methodically by using a good "hill climbing" algorithm,[1] but the essential idea is to adjust the guess of A until no smaller error can be found.

However, if the system being modeled has equation noise $W(t) \neq 0$, then the naive simulation method may give minimum error J for a completely wrong value of A, since the real system, under the cumulative impact of random disturbances, $W(t)$, may "drift" away from the deterministic, $W(t) = 0$ trajectory, as shown, for example, by Forrester (1961, appendix K). In fact, Peterson (1976) has shown that, for such noise-driven systems, the naive simulation method in effect ignores most of the data.

Ordinary Least Squares When driving noise $W(t)$ is present, but measurement error $V(t)$ is absent, the modeler can in general obtain better estimates of A by reinitializing the system at each data point, and then applying the same squared residual error function J as in the naive simulation method. The new iteration is called ordinary least squares, *OLS*:

$$\hat{X}(t) = A*Z(t-1)$$

$$\hat{Z}(t) = \hat{X}(t),$$

as illustrated in figure 11.2. The dots are the data $Z(t)$, the dashed lines are the simulation trajectories between data points, and the vertical bars are the residuals. Whenever the simulation reaches a data-sample time, the system is reinitialized, so that each segment of the simulation begins

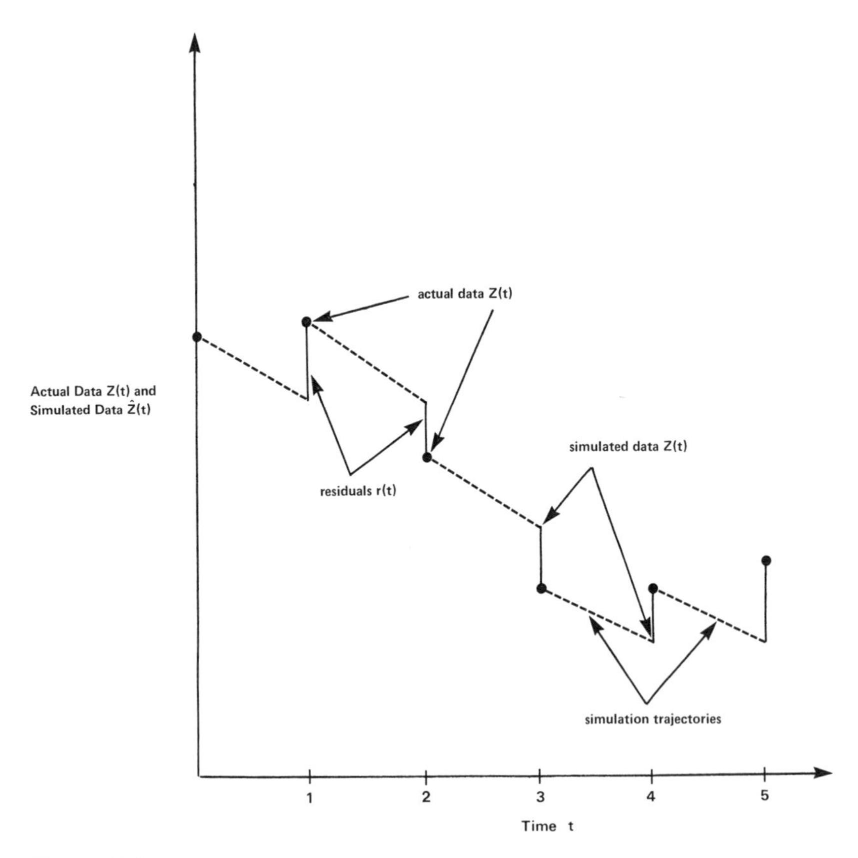

Figure 11.2
Noise-driven system and accurate data with *OLS* estimates

on a data point. As in the other methods the simulation is repeated with different values of A, and the value of A that yields the minimum error J is taken as the estimate, A. (Note that *OLS* estimates are seldom actually computed this way. By taking advantage of the assumed perfection of data, $V(t) = 0$, it is possible to compute *OLS* using a computational shortcut that avoids simulation altogether. Either computation yields the same *OLS* estimates, but the method described here clarifies the relationship between *OLS* and the other methods).

Wonnacott (1970) has shown that the *OLS* method tends toward a good estimate of A, so long as $V(t) = 0$. But the method breaks down with measurement errors, leading to grossly inaccurate estimates of parameters in a typical system dynamics model. The intuitive reason for the failure is as follows. The motivation for reinitializing the model at each data point is to keep the simulation close to the true state of the system, so that any divergence in behavior, as measured by the residuals, would be meaningful. But reinitialization in the presence of noisy data is not likely to keep the simulated state close to the real state. Large residuals may emerge from even a perfect model under *OLS*.

Full-Information Maximum Likelihood The essence of *FIMLOF* is to reinitialize the system at each data point, at the value of $X(t)$ where the system is most likely to be given all available data. To do so, the simulation must compute not only the predicted state at each data point but also the size of the expected error (standard deviation) of the predicted state. Therefore *FIMLOF* uses the iteration:

$$\hat{X}(t/t-1) = A*\hat{X}(t-1/t-1)$$

$$\hat{Z}(t/t-1) = \hat{X}(t/t-1),$$

where $\hat{X}(t/t-1)$ is the most likely value of $X(t)$, given all information through time $t-1$, $\hat{X}(t-1/t-1)$ is the most likely value of $X(t-1)$, given the same information, and $\hat{Z}(t/t-1)$ is the best guess of the next measurement $Z(t)$, given all the previous data $Z(0) \ldots Z(t-1)$.

The simulation is then updated at $\hat{X}(t/t)$, which is defined as the most likely value of $X(t)$, given all information through time t. This information is embodied in $Z(t)$, in $\hat{X}(t/t-1)$, and in the variances of these two quantities. The variance of $Z(t)$ is simply the variance of the process $V(t)$. The variance of $\hat{X}(t/t-1)$ is automatically derived from the variances of the processes $V(t)$ and $W(t)$, from the variance of the guess of the initial conditions, $\hat{X}(0/0)$, and from the structure of the model. The computation of the variances is not detailed here; the equations which compute them constitute an "optimal filter," which is documented by Schweppe (1973) and Kalman (1960), and presented in the appendix of this paper.

Variances are employed in *FIMLOF* to avoid the pitfalls observed in *OLS* and *NS*. The process is illustrated in figure 11.3. As in *NS* and *OLS*, the initial conditions of the model are based on the first data point. The system is simulated to the time of the first data point,and the first residual is computed as

$$r(1/0) = Z(1) - \hat{Z}(1/0).$$

So far, the process has been the same as for *NS* and *OLS*. The difference lies in how the model is reinitialized at the data point $Z(1)$. Instead of leaving the model state at $\hat{Z}(1/0)$, as in *NS*, or adjusting the model state

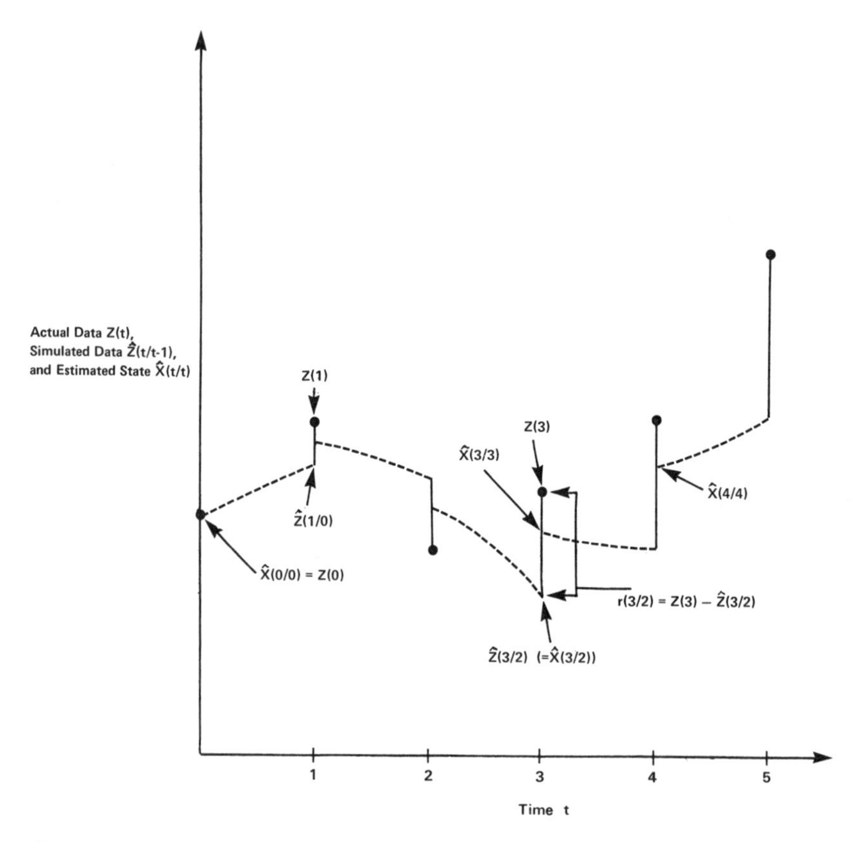

Figure 11.3
Noise-driven system and noisy data with *FIMLOF* estimates

completely to match $Z(1)$, as in *OLS*, *FIMLOF* reinitializes the model to a compromise point somewhere *between* $\hat{Z}(1/0)$ and $Z(1)$. The compromise is based on the variances of $\hat{Z}(1/0)$ and $Z(1)$. If the variance of $Z(1)$ is large (noisy measurements), but the variance of $\hat{Z}(1/0)$ is small (little driving noise $W(t)$), then the reinitialization will be close to $\hat{Z}(1/0)$, as in *NS*. If, at the other extreme, the variance of $Z(1)$ is small, but the variance of $\hat{Z}(1/0)$ is large (accurate data but highly uncertain model), then the model will be reinitialized close to $Z(1)$, as in *OLS*. In general, the reinitialization in *FIMLOF* will be somewhere between the two extremes, at a point always chosen as the most likely value of the model state, given the available data. Schweppe (1973) has shown that, with this optimal reinitialization (called "optimal filtering" in control theory) and a loss function based on the residuals $r(t/t-1)$, parameter estimates will be those that are most likely, given all the information contained in the data and model structure. In fact, the *FIMLOF* loss function (see the appendix) is the negative logarithm of the *likelihood* (probability) that the observed data could have been produced by the assumed model (including the guesses of the parameters). By making astute guesses of the parameters (again, automatically, according to a "hill-climbing" algorithm, as in Murray, 1972), the modeler arrives at the desired "maximum likelihood" estimated.

Feasibility The likelihood computation is exact only in the case of linear systems with Gaussian noise $W(t)$ and $V(t)$. But the approximate likelihoods computed by nonlinear filters, as in the GPSIE software package, are remarkably accurate. Experiments on nonlinear engineering models by Moore (1972), Mehra (1973), and Arthur (1976) have indicated that nonlinearities are not a serious problem.

FIMLOF (as implemented in GPSIE) has also been tested on system dynamics models of social systems. Table 11.1 summarizes the results of one such experiment, chosen to facilitate comparison with Senge (1974). Senge estimated parameters from noisy simulation data generated by a nonlinear, dynamic model of market growth, as published by Forrester (1968). The model consists of nine dynamic difference equations (defin-

Table 11.1
Comparison of estimation techniques for ninth-order nonlinear system with errors in variable

Parameter name	True value	GPSIE estimate	*OLS* estimate
SEM	400	392	4349
*SED*1	−0.0281	−0.29	−0.430
*SED*2	−0.0295	−0.0295	0.096
*SED*3	0.00228	0.00228	−0.0074
*PCF*1	0.61782	0.615	3.7117
*PCF*2	−0.13244	−0.132	−0.74891
*CEF*1	−0.0698	−0.0693	0.03966
*CEF*2	0.12442	0.1245	−0.14609
*CEF*3	−0.08138	−0.0813	0.13953
*CEF*4	0.027704	0.02704	−0.03144
DRAT	1	0.97	1.3
SAT	20	19.85	18.5

ing nine "level" variables). Senge simulated the model, using random-number generators to introduce both equation errors and measurement errors in different amounts. In the particular experiment compared below, Senge introduced equation errors ranging from 6 percent to 60 percent of the mean of the endogenous variables, and obtained excellent estimates of the thirteen system parameters, using *OLS* and *GLS*. However, when Senge introduced 10 percent measurement error, he obtained large errors in the parameter estimates. Table 11.1 shows the results Senge obtained, compared with the estimates obtained with the *FIMLOF* software package GPSIE under the same conditions. The results indicate that *FIMLOF* techniques, as implemented in GPSIE, may yield accurate results, even in the presence of system nonlinearities and measurement error which may cause difficulties with simpler estimation techniques.

In general, for nonlinear models of the system dynamics type, the approximations involved in *FIMLOF* cause errors no more serious than those due to other approximations in computer simulation, such as numerical-integration error and round-off error.

Although seemingly lengthy and indirect, the repeated guess-and-simulate iteration of *FIMLOF* allows great flexibility and reliability. For example, nowhere does the method require that all the state variables be measured, or the availability of data at each time step, or even that the data be distributed at constant intervals. If a data point is missing, for example, the simulation simply continues to the time of the next valid data point. Residuals are computed only at data points, not necessarily at all model time steps.

Similarly, the mathematics of *FIMLOF* can deal with unknown (unmeasured) exogenous inputs, cross-sectional data, short data series, nonwhite noise, and such indirect measurements as yearly summations, averages, and functions of multiple state variables.

The alert reader may object that *FIMLOF* computations require the variances of measurement errors $V(t)$ and equation errors $W(t)$, which are seldom known with confidence. One of the strengths of *FIMLOF* is that the variances can be simply included in the list of parameters to be estimated. *FIMLOF* thus yields not only the most likely values of the model parameters but also the most likely magnitudes of equation errors and measurement noise.

The features of the *FIMLOF* method are summarized in table 11.2. *FIMLOF* is compatible with the characteristics commonly found in system dynamics models.

Table 11.2
Conditions for applying *FIMLOF* (as implemented in GPSIE)

1. Nonlinearities in model structure
2. Nonlinear measurement functions
3. Measurement error (errors in variables)
4. Mixed sampling intervals (can, for example, estimate a weekly model, using monthly and yearly data)
5. Missing data (without sacrificing other data at the same sample time)
6. Models with unmeasured endogenous variables
7. Cross-sectional, time-series endogenous variables
8. Unknown characteristics of equation errors and measurement noise.

Structural Estimation

The preceding observations on parameters and nonlinear systems raise other interesting possibilities. In fact, for nonlinear systems, we must redefine "structure," as opposed to "parameter." For time-invariant linear systems, the distinction is clear: the system must, by definition, take the form

$$\mathbf{x}(n) = \mathbf{A}\mathbf{x}(n-1) + \mathbf{B}\mathbf{u}(n) + \mathbf{w}(n),$$

where $\mathbf{A}$ and $\mathbf{B}$ are constant matrices, $\mathbf{x}(n)$ is the state vector at time n, $\mathbf{u}(n)$ is a vector of known inputs, and $\mathbf{w}(n)$ is a white normal process of mean 0 and covariance $\mathbf{Q}$. In this case, the parameters are simply the constant coefficients of the matrices $\mathbf{A}$, $\mathbf{B}$, and $\mathbf{Q}$. A similar definition can be made for systems that are linear in the parameters. For example, in the system

$$\mathbf{y} = \mathbf{X}\mathbf{b},$$

where $\mathbf{y}$ is a vector of outputs, $\mathbf{X}$ is a matrix of variables that may be functions of exogenous inputs and of lagged values of $\mathbf{y}$, and $\mathbf{b}$ is defined as the vector of parameters.

General nonlinear systems, however, will require a more general definition of parameters. By definition, a *parameter* in a nonlinear system is a constant exogenous input to the system. It follows that a parameter in a nonlinear system may enter the system in any nonlinear fashion. The parameter may be known or unknown, but it is always a constant whose value is not determined by the rest of the system.

This seemingly straightforward definition has the following further implication: *in nonlinear systems, parameters may take on qualities usually associated with structure.* For example, consider the system

$$\mathbf{x}(t) = \theta\mathbf{f}(\mathbf{x}(t-1)) + (1-\theta)\mathbf{g}(\mathbf{x}(t-1)).$$

By any reasonable definition, θ would be considered a parameter in the equation. But θ determines the structure of the system. If $\theta = 0$, then the system has the structure determined by the function **f**; if $\theta = 1$, then the system structure is determined by the function **g**.

Therefore care must be taken in applying the usual connotations to the terms "structure" and "parameter" when dealing with nonlinear systems. By model building such as the example just illustrated, the modeler may estimate parameters which de facto result in the estimation of structure.

The estimation of structure (as in estimating θ in the above system) may be thought of as a kind of continuous hypothesis test. The maximum-likelihood value of θ may be thought of as selecting the most likely structure from the range of structures implied in the equation. In addition, completely separate models may be compared by computing the likelihood of each with respect to the same data base.

Neither parameters nor structure can be usefully estimated from mere numerical data and thin air. Estimation inherently entails a choice from a range of alternatives. A well-hypothesized model should define a range of plausible alternatives consistent with the purposes of the study at hand. Since *FIMLOF* imposes no hard constraints on the form or content of the model, the range of plausible alternatives is limited only by the imagination and resources of the model-builder.

Validity and Consistency Tests

The essence of validity is that the model be consistent with all available information, in the context of the purpose of the model. The automatic consistency tests related to *FIMLOF* represent a subset of this general notion of validity. The *FIMLOF*-based tests measure the consistency of the model with available *numerical* data. However, qualitative knowledge can usually be quantified to some approximation, and *FIMLOF* can make use of approximate data.

The consistency tests of *FIMLOF* come in two kinds. First, the likelihood evaluations for each set of parameter guesses provide information, and, second, the optimal filter itself provides several internal consistency measures. Most of these consistency tests are based on mathematical derivations of proprieties which the likelihood evaluations and filter outputs must have if the model is an accurate representation of the real system that actually produced the data. If the properties are not observed in the *FIMLOF* output, then there is some inconsistency between model and data.

Use of the Likelihood Surface The "loss function" computed in *FIMLOF* is the negative natural logarithm of the likelihood that the data could have been generated by the model. For each different model or set of parameter guesses, the same data will yield, in general, a different "log likelihood." Therefore the data and model define a surface over the space of all possible parameter values. One property of the log-likelihood surface is especially useful in interpreting parameter estimates. At the global maximum of the log-likelihood surface, the curvature of the surface is a measure of the quantity of information about the unknown parameters contained in the data. In the extreme, if the likelihood surface is flat (so that all parameter values are equally likely), then there is no information in the data with respect to the model and parameters being estimated. Similarly, if the surface curves sharply downward from the maximum, then the estimates are highly precise, and the data (however noisy) contain a great deal of information about the model. More precisely, the second derivative of the log-likelihood surface with respect to parameter A is the variance of the uncertainty in the estimate of A.

Confidence Tests from the Optimal Filter The optimal filter computes not only the residuals $r(t/t-1)$ but also what the standard deviation of the residuals should be, if the model is correct. The residuals, when normalized by their theoretical standard deviations, can be shown (Schweppe, 1973) to have two properties. First, the normalized residuals should have a constant variance of one; second, the sequence of residuals should be a white process. Since these properties of the residuals process are not employed directly in maximizing the log likelihood, they provide an independent test of model validity. Furthermore, experience shows that residual-based tests are sensitive to small errors in model specification.

The two theoretical properties of the residuals—whiteness and unit variance—provide two kinds of consistency tests. First, the whiteness may be tested by computing correlation measures of the normalized residuals. Each residual (in the case of multiple-dimensional measurements) should have a correlation coefficient of one with respect to itself, zero serial correlation with respect to lagged values of itself, and zero cross correlation with all other residual processes (see the appendix for mathematical definitions). The correlation test of the residuals not only indicates whether the model is consistent with the data, it may also reveal what is wrong if the test fails. For example, the pattern of serial correlation coefficients may reveal the first-order time constant associated with a delay missing from the model structure.

Second, the theoretical unit variance of the residual processes provides a test of internal consistency. The log-likelihood function consists of two terms: (1) a sum of squared residuals term, *SUMSQ*, which is analogous to the *OLS* loss function, and (2) a term which is independent of the size of the residuals. The expected magnitude and standard deviation of the *SUMSQ* term at the true parameter values can be predicted ahead of time. If the actual *SUMSQ* differs from the predicted range, then either the model is inconsistent with the data, or the global maximum of the log-likelihood function has not yet been found.

Detection of Bad Data

The preceding discussion explained how noisy approximate data can be used to help estimate and evaluate system structures. But most collections of data contain some points so much in error as to be best deleted. Such "bad data" points may arise from typographical errors, improper accounting procedures, and other gross malfunctions of data collection. Instead of containing useful information about the system, the bad data points serve only to mislead.

The *FIMLOF* methodology includes convenient techniques for detecting and isolating bad data. The basic idea behind the techniques is to look for residuals that are clearly too large. The obvious difficulty is how to define "too big." The answer is provided by the optimal filter, which computes the expected standard deviation of each residual under the assumption that the model and data are consistent. As a matter of practicality, a bad data point will create large residuals even if the model is still somewhat approximate. Therefore, even if the model has been estimated using data which contains bad data points, a residual more than four or five standard deviations away from its expected value can be taken to indicate a bad data point.

A complication arises when, as is usual, more than one variable is measured at the same time. In such a case, a large residual still reveals the presence of a bad data point, but does not necessarily indicate which of the several measurements is at fault. However, the optimal filter of the *FIMLOF* method provides the information required to decide which component is in error. From the variances and covariances by the filter, the evaluation can compute "normalized updated" residuals (see Peterson, 1975) which pinpoint the bad data points in both time and space.

Although bad data can often be spotted by visually scanning graphed data, the *FIMLOF* techniques are useful for two reasons. First, the techniques can be completely automated, allowing a thorough checking of large data files. Second, bad data is not always readily apparent, even when the data is presented in graphical form. The "wrongness" of a data point is often seen only in the context of the dynamic structure behind the data, as well as measurements of other variables. Preliminary experience (Peterson, 1975) indicates that bad data may be uncovered quickly using the *FIMLOF* methods, even in data files which have been manually inspected for errors.

Estimation of the System State

A useful by-product of the optimal filter is the computation of the most likely state of the system at a given time. The estimated state can be used to initialize the system for forecasting. The estimated state may also yield insight, or aid in decision making. For example, a decision-maker would like to know which inputs are limiting a production process, or in which region of a nonlinear function the system is operating. The filter provides not only an estimate of the true state of the system but also gives confidence bounds, by way of the variances and covariances of the state-variable estimates. Such computations can also be validly continued into the future.

Confidence Bounds for Forecasts

Simulation models are often used to predict the future evolution of a system. Usually, the model is initialized at some approximation of present conditions, and the (deterministic) simulated trajectory is taken as a best guess of the future. Such a forecast may be inaccurate for three reasons. First, of course, the model structure may be inaccurate. But even if the model is "perfect," two sources of uncertainty may bring about inaccurate forecasting. To the extent that the real system is driven by uncertain processes (events modeled as random), the future evolution of the real system is likely to drift away from any computed trajectory, thereby expanding the frequency and magnitude of errors in the forecast. Finally, there is the difficulty of deciding what initial conditions to use for the forecasting simulation. Since many state variables in a model may be unmeasured, the most likely present state of the system may not be obvious from a casual inspection of the data.

Forrester (1961, app. K) discusses these ideas qualitatively, illustrating the interaction among model accuracy, knowledge of the present state of the real system, and forecast accuracy. The *FIMLOF* techniques allow these factors to be assessed quantitatively. Forecasts computed by the filter include not only the "expected" future trajectory of the system but also confidence bounds on the trajectory. The "initial conditions" of confidence bounds are derived from the computed variance of the previously explained present state estimate. The filter then computes the a priori variance of future state estimates as a function of the initial variance, the model structure, and the variances of the random system inputs.

The confidence bounds on the forecast clearly show to what extent, and how far into the future, a given model may be expected to yield accurate predictions. In many social systems, the confidence bounds diverge rapidly as the simulation extends farther ahead in time. The timing and severity of the divergence will depend on the state of the model, its structural accuracy, the nature of the structure, the accuracy of the initial conditions, and the severity of random inputs to the system.

11.2 GPSIE Implementation

The various *FIMLOF*-based techniques discussed here have been implemented in the GPSIE computer program. GPSIE is a large precompiled program that links with a user-written program describing the particular model of interest. The resulting package can be used to load data, compute likelihoods via optimal filtering, search for maxima in the likelihood function, compute the validity statistics discussed here, and plot the results. GPSIE embodies a large number of options for dealing with special cases and for maintaining efficient computation in various circumstances. (For more details on GPSIE, see Peterson, 1974 and 1975).

An obvious concern with iterative methods and statistical analyses is the threat of high computation costs. GPSIE, for example, imposes no inherent limits on the model size or data base, but requirements of computer time or storage may obviously become extravagant for large systems. For example, the cost of parameter estimation and validity tests for a tenth-order system with 1,000 data points would typically fall between $200 and $300 on a large time-sharing system.

The computational costs of some *FIMLOF* computations vary with the cube of the system dimensions. Therefore it often pays to break the model into sectors for individual analysis. The sectors may then be recombined for final validity testing, requiring but a single filtering computation with the entire system intact.

11.3 Conclusions

There are two implications of the *FIMLOF* techniques for the field of system dynamics. First, the techniques may increase the efficiency and quality of system dynamics modeling by complementing the manual simulate-analyze-correct techniques. Parameter estimation, consistency tests, and confidence bounds may efficiently indicate areas of sensitivity or inconsistency which might otherwise be found with difficulty. A failed consistency test or unreasonable estimated parameter, furthermore, may not only reveal the presence of trouble but also may suggest an appropriate remedy in model structure.

Second, the *FIMLOF* techniques may extend the practice of system dynamics into additional fields and disciplines. As techniques such as *FIMLOF* become more widely understood and available, system dynamics type of models may be employed more often for data-related activities, including forecasting, data validation, and performance monitoring.

Note

1. Also called unconstrained optimalization. See Murray (1972) for examples.

References

Arthur, William B. 1976. Private communication.

Forrester, Jay W. 1961. *Industrial Dynamics.* Cambridge, Mass.: MIT Press.

Forrester, Jay W. 1968. "Market Growth As Influenced By Capital Investment." *Industrial Management Review,* vol. 9 (Winter), pp. 83–105.

Hamilton, H. R. et al. 1969. *System Simulation For Regional Analysis.* Cambridge, Mass.: MIT Press.

Kalman, R. E. 1960. "A New Approach To Linear Filtering and Prediction Problems." *Journal of Basic Engineering,* series D, vol. 82. (March), pp. 35–45.

Mehra, R. K., and J. S. Tyler. 1973. "Case Studies in Aircraft Parameter Identification." In *Identification and System Parameter Estimation* edited by P. Eykhoff. New York: American Elsevier Publishing Co.

Moore, R., and F. C. Schweppe. 1972. "Adaptive Control for Nuclear Power Plant Loan Changes." Proceedings of the Fifth World Congress of the International Federation of Automatic Control, Paris.

Murray, W. 1972. *Numerical Methods For Unconstrained Optimization.* New York: Academic Press.

Peterson, D. W., and F. C. Schweppe. 1974. "Code For a General Purpose System Identifier and Evaluator: GPSIE." *IEEE Transactions on Automatic Control,* vol. Ac–19 (December), pp. 852–854.

Peterson, D. W. 1975. "Hypothesis, Estimation, and Validation of Dynamic Social Models." Ph.D. dissertation. Department of Electrical Engineering, MIT, Cambridge, Mass.

Peterson, D. W. 1976. "Parameter Estimation for System Dynamics Models." Proceedings of the Summer Computer Simulation Conference, Washington, D.C.

Runge, D. 1976. "Labor-Market Dynamics: An Analysis of Mobility and Wages." Ph.D. dissertation. Alfred P. Sloan School of Management, MIT, Cambridge, Mass.

Schweppe, F. C. 1973. *Uncertain Dynamic Systems.* Englewood Cliffs, N.J.: Prentice-Hall.

Senge, P. M. 1974. "Evaluating the Validity of Econometric Methods for Estimation and Testing of Dynamic Systems." System Dynamics Group memo of February 12 D–1944–2. Alfred P. Sloan School of Management, MIT, Cambridge, Mass.

Senge, P. M. 1974. "An Experimental Evaluation of Generalized Least Squares Estimation." System Dynamics Group memo of November 14 D–1944–6. Alfred P. Sloan School of Management, MIT, Cambridge, Mass.

Wonnacott, P. J., and T. H. Wonnacott. *Econometrics.* New York: John Wiley and Sons.

Appendix: Mathematical Definitions

This appendix is for the benefit of the mathematically inclined reader. It summarizes the equations of the *FIMLOF* techniques discussed qualitatively in the text. The first section defines the notation of the model, its relationship to data, its uncertainty, and the linearization of the model; the second gives the equations of the optimal filter and the accompanying likelihood calculation; the third gives the equations for two of the confidence tests discussed in the text; and finally, the fourth summarizes the techniques of bad data detection. For more details on bad data detection, see Peterson (1975).

System Dynamics and Linearization

State equations:
$\mathbf{x}(n) = \mathbf{f}[x(n-1), \mathbf{u}(n), \mathbf{w}(n), n].$

Measurement equations:
$\mathbf{z}(n) = \mathbf{h}[\mathbf{x}(n), \mathbf{v}(n), n].$

Index of data samples:
$n = 1, 2, \ldots N.$

Initial conditions:
$\mathbf{x}(0) = N[\mathbf{x}_0, \Psi]^*.$

Equation errors (driving noise):
$\mathbf{w}(n) = N[\mathbf{0}, \mathbf{Q}(n)]^*.$

Measurement errors:
$\mathbf{v}(n) = N[\mathbf{0}, \mathbf{R}(n)]^*.$

Linearization about estimated state:

$$\tilde{\mathbf{F}}(n) = \frac{\partial \mathbf{f}}{\partial \mathbf{x}}\Big|_{\mathbf{x} = \hat{\mathbf{x}}(n-1 \mid n-1)},$$

$$\tilde{\mathbf{H}}(n) = \frac{\partial \mathbf{h}}{\partial \mathbf{x}}\Big|_{\mathbf{x} = \hat{\mathbf{x}}(n \mid n-1)},$$

$$\tilde{\mathbf{Q}}(n) = \left(\frac{\partial \mathbf{f}}{\partial \mathbf{w}}\right) \mathbf{Q}(n) \left(\frac{\partial \mathbf{f}}{\partial w}\right)' \Big|_{\mathbf{w} = \mathbf{0}},$$

* $N[\mathbf{m},\mathbf{c}]$ denotes a normal, white process with mean **m** and covariance matric **c**.

$$\tilde{\mathbf{R}}(n) = \left(\frac{\partial \mathbf{h}}{\partial \mathbf{v}}\right) \mathbf{R}(n) \left(\frac{\partial \mathbf{h}}{\partial \mathbf{v}}\right)' \Big|_{\mathbf{v}=\mathbf{0}}$$

Filter Equations

Predicted state:
$\hat{\mathbf{x}}(n|n-1) = \mathbf{f}[\hat{\mathbf{x}}(n-1|n-1), \mathbf{u}(n), \mathbf{0}, n].$

Predicted measurement:
$\hat{\mathbf{z}}(n|n-1) = \mathbf{h}[\hat{\mathbf{x}}(n|n-1), \mathbf{0}, n].$

Residuals:
$\delta_z(n|n-1) = \mathbf{z}(n) - \hat{\mathbf{z}}(n|n-1).$

Predicted state covariance:
$\sum_x(n|n-1) = \tilde{\mathbf{F}}(n) \sum_x(n-1|n-1)\tilde{\mathbf{F}}'(n) + \tilde{\mathbf{Q}}(n).$

Predicted measurement covariance:
$\sum_z(n|n-1) = \tilde{\mathbf{H}}(n) \sum_x(n|n-1)\tilde{\mathbf{H}}'(n) + \tilde{\mathbf{R}}(n).$

Normalized predicted measurement residuals:
$\tilde{\delta}_z(n|n-1) = \sqrt{\sum_z(n|n-1)}^{\,-1} \delta_z(n|n-1).$

Updated state covariance:
$\sum_x(n|n) = \left[\sum_x^{-1}(n|n-1) + \tilde{\mathbf{H}}'(n)\tilde{\mathbf{R}}^{-1}(n)\tilde{\mathbf{H}}(n)\right]^{-1}.$

Filter Gain:
$\mathbf{K}(n) = \sum_x(n|n-1)\tilde{\mathbf{H}}'(n) \sum_z^{-1}(n|n-1).$

Updated state estimate:
$\hat{\mathbf{x}}(n|n) = \hat{\mathbf{x}}(n|n-1) + \mathbf{K}(n)\delta_z(n|n-1).$

Log likelihood:
$\xi(n) = \xi(n-1) - \frac{1}{2}\tilde{\delta}_z'(n|n-1)\tilde{\delta}_z(n|n-1) - \frac{1}{2}\ln \left|\sum_z(n|n-1)\right|.$

Initial conditions:
$\hat{\mathbf{x}}(0|0) = \mathbf{x}_0, \sum_x(0|0) = \Psi, \xi(0) = 0.$

Confidence Tests

Residual correlation matrices:

$$\mathbf{R}(j) = \frac{1}{N-j} \sum_{n=1}^{N-j} \tilde{\delta}_z(n|n-1)\, \tilde{\delta}_z'(n+j|n+j-1).$$

If model and data are inconsistent, then $R(0) \approx I$ (identity matrix), and $R(\mathrm{j}) \approx 0$, $\mathrm{j} \neq 0$.

SUMSQ statistic:

$$SUMSQ = \sum_{n=1}^{N} \tilde{\delta}'_z(n \mid n-1)\tilde{\delta}_z(n \mid n-1).$$

The expected values of *SUMSQ* is equal to the total number of scalar data points minus the number of unknown parameters. The standard deviation of *SUMSQ* is equal to the square root of twice its expected value.

Bad Data Detection

Normalized updated measurement residuals (*NUMR*):

$$\mathbf{r}_z(n \mid n) = \left[\text{diag} \quad \Sigma^{-1}_{z}(n \mid n-1) \quad \right]^{-\frac{1}{2}} \tilde{\mathbf{R}}^{-1} \ (n)[\mathbf{z}(n) - \hat{\mathbf{Z}}(n \mid n)].$$

Normalized updated state residuals (*NUSR*):

$$\mathbf{r}_x(n \mid n) = \left[\text{diag} \quad \tilde{\mathbf{H}}'(n) \Sigma^{-1}_{z}(n \mid n-1)\tilde{\mathbf{H}}(n) \quad \right]^{-\frac{1}{2}} \Sigma^{-1}_{x}(n \mid n-1)$$
$$[\hat{\mathbf{x}}(n \mid n-1) - \hat{\mathbf{x}}(n \mid n)].$$

The *NUMR* and *NUSR* processes interact with each other and must be considered together. They have two useful properties. First, both *NUMR* and *NUSR* have constant unit variance. That is, each component of *NUMR* and *NUSR* has a constant standard deviation of 1, under all circumstances, as long as the model is valid and the data conforms to the model. Therefore one or more components of *NUMR* and *NUSR* with absolute value greater than 3 or 4 is a reliable sign of a bad data point somewhere at the corresponding sample time.

Second, at a sample time involving a bad data point, the component of *NUMR* or *NUSR* with the maximum absolute value corresponds to the component of $\mathbf{z}(n)$ or $\mathbf{x}(n)$ which is in error as shown by Peterson, (1975). For example, a typographical error in the first component of $\mathbf{z}(3)$ may, depending on the model structure, cause several components of both $\mathbf{r}_z(n/n)$ and $\mathbf{r}_x(n/n)$ to exceed the acceptable limit (for example, 4). But the component with the largest absolute value identifies the specific component of $\mathbf{z}(3)$ or $\mathbf{x}(3)$ which contains the typographical error. (In the case of $\mathbf{x}(n)$, the error might be in an exogenous input to the equation determining the component of $\mathbf{x}(n)$). These properties of *NUMR* and *NUSR* permit the creation of computer programs that automatically identify and delete bad data points, and the efficient screening of data sets for questionable entries. As shown by Peterson (1975), they can also be helpful in model validation and model improvement.

Part VI
Implementation

Managerial Sketches of the Steps of Modeling

12

Jennifer M. Robinson

12.1 Introduction

The roads by which men arrive at their insights into celestial matters seem to me almost as worthy of wonder as those matters in themselves. —Johannes Kepler (1609)

The modeling literature tends to refer to models as abstract entities: theory, method, data, and behavior are discussed at length; reference to the modeler, the client, and the institutional context in which they interact is almost religiously avoided. The minds that analyze social and economic processes are seldom found reflecting on the social processes and economic forces which affect them personally as members of a profession.

While understandable in the context of scientific research, such avoidance of the human and social side of modeling is misleading. Models are not brought by storks. They don't grow up in the realm of pure reason, and they aren't nurtured by air. Like another product that storks don't bring, models are conceived in human interaction, are shaped by the environment in which they are formed, and require time, money, and patience to grow into anything useful to society. The practical outcome of a modeling study is as likely to be influenced by the conditions of its conception, the environment in which it is formed, and the amount of love bestowed on the project as by the data the model contains or the theory on which it is built. A well-managed model that is mediocre by academic standards may have a far greater and more productive impact than a technically superlative model that no one wanted.

This chapter grew out of field observations of ten public-sector models. It rests on the following assumptions:

1. Certain managerial difficulties are inherent in the modeling process.

These difficulties beset all modelers and render many modeling studies ineffectual.
2. Greater awareness of the modeling process can improve management of these inherent difficulties.
3. Process awareness may be developed through detailed observation of actual modeling projects.
4. A self-critical mode is encouraged by presenting (blamelessly, or better yet, humorously), material with which the reader can identify.

Working from these premises, I have grouped the actions that occur during the lifecycle of a model into seven managerial steps: preconceptions, establishing contact, conceptualization, construction, testing, documentation, and implementation. In the course of the chapter I shall caricature what typically does, but should not, happen at each of these steps, and suggest ways of avoiding the normal snags and bogs. For simplicity, I shall discuss the steps in straightforward, one-thing-happens-at-a-time fashion. The illusion that modeling proceeds so tidily has been dispelled elsewhere (Randers, 1972, and Hammond, 1974). Here we can make do with a reminder that some steps (particularly conceptualization) tend to be iterative, while others, such as testing and implementation, tend to be ongoing activities.

Most model management problems originate in the early steps of modeling, but their bad effects manifest themselves in later stages. Thus my recommendations, particularly for testing and implementation, will frequently refer to what should have been done in earlier steps.

12.2 Preconceptions

Thus we economists and other social scientists are now studying intensively how people behave, and how they are motivated and then conditioned both by their inherited constitution and by their environment. . . . Only about the peculiar behavior of our own profession do we choose to remain naive. . . . The point is that we could better avoid biases, and could therefore expect more rapid progress in the social sciences, if we are a little less naive about ourselves and our motivations. A minimal desideratum is that we be always aware of the problem and attain some degree of sophistication abouth the operation of the personal and social conditioning of our research activity.—Myrdal (1968)

Preconceptions are the set of generally unspoken beliefs, assumptions, and expectations that surround the initiation of a modeling effort. In other words, preconceptions are the often implicit and unrecognized forces that shape the model before it becomes explicit, and continue to operate behind the model throughout its development. To an extent, preconceptions are adaptive and necessary. Without them, modeling could not begin because there would be no ordering principles. Yet, sometimes preconceptions are maladaptive. That is, they may order the perceptions of modeler and client in a manner that poorly explains the real-world situation and poorly answers the client's needs.

Preconceptions take many forms. Some are firmly entrenched, such as the formalized preconceptions of a modeling paradigm, a strong belief in the free market system, or a general skepticism toward, or overenthusiasm for, modeling in general. Others are ephemeral and may fade or change during the course of building the model. The modeler may have just read a report on changing global weather patterns and be convinced for awhile that they must be included in the model. The client may have recently received a congressional mandate to pay more attention to income distribution, but over the course of model construction his focus of attention may shift back toward balance of payment problems. Be they deeply entrenched beliefs or thoughts of the hour, modeler and client preconceptions strongly influence the modeling process.

Modeler's Preconceptions

Professional preconceptions, particularly the modeling paradigm, will shape the modeler's methodology and structure his thinking. They will cause him to focus his attention on certain problems and may make him unwilling to consider others. The system dynamicist's perspective, for example, biases him toward aggregate, closed-loop structures. These biases are apt to blind him to (1) dispersion problems such as geographical variations and income distribution, (2) hierarchical structures such as differences in degrees and forms of centralization, and (3) parameter-determined short-term behavior (see Meadows, chapter 2).

Unless we can devise some demon to sit atop preconceptions and sort out the maladaptive, we are going to have to make do with primitive common sense. Here common sense would say "be aware and act with awareness." Preconceptions are always there. To the extent that the modeler is aware of them, he can make them explicit and deal with them directly. To the extent that he does not see them, preconceptions will lurk behind his work, shaping it in fashions unrelated to his intended purpose.

More specifically, be self-critical, examine your motives and your a priori assumptions. Look for the biases inherent in your techniques.

Continually ask what you are trying to do and why. Introspection and self-correction are the negative feedback that keep the modeling study on the course of its goals. Without introspection, there is little to prevent preconceptions from steering the modeling study into irrelevance.

The usual training of modelers does little to encourage self-critical awareness. Moreover, the technical nature of the field may actively select against the "soft" introspective attributes that counteract the bullying of preconceptions. The full antidote to preconception problems requires restructuring of values in the modeling profession.

Client's Preconceptions

In contrast to the modeler's paradigm-centered preconceptions, client preconceptions often lack theoretical structure and are highly responsive to the institutional environment. Rightfully so, for the client can seldom affect change without cooperation from other members of his institution and its supporters. A client from the U.S. Agency for International Development may go out of his way to maintain the good will of Congress; the World Bank may not publish reports that would embarrass member nations. Any client may go out of his way to respect the opinion of a person or group he considers important.

Changing institutional considerations can be very disruptive to the modeling process. As an extreme example, one agricultural model that was originally meant to focus on Nigeria was moved to Korea when a revolution blocked the Nigeria study.[1] A more common occurrence would be that of a client commissioning a population study when the public is concerned about population, and then ignoring it as agency interest shifts from population to resource allocation.

Be conscious of client preconceptions. Imagine yourself in the client's place. Sense the institutional pressures the client is under. Those pressures will be transmitted to bear on you. Distinguish clearly between the client's personal biases and the institutional pressures he is under. Ally yourself with the client in counteracting institutional pressures such as bureaucratic fragmentation and reductionist approaches that are detrimental to holistic problem solving. Don't accept client biases that will block model usefulness (for example, short-term focus, point-predictive expectations), but don't alienate the clients either. You cannot affect a person's preconceptions unless he respects and trusts you. Work on gaining confidence. Once confidence is gained, begin on key biases slowly. Most of the biases that repeatedly confront system dynamicists stem from short-term, reductionist static thinking. These biases are best overcome by teaching the principals of dynamic systems. Your main task thus is an affirmative one—teaching systems logic. Don't get so embattled in the defense of details that you forget to convey the main point.

12.3 Establishing Contact

Modeler-client contact begins when the specific modeling project is first discussed by the modeler and client, and continues until a contract is agreed upon. The more mundane decisions of model building are heavily concentrated in this early step. Financial arrangements are established, as well as deadlines and reporting schedules, physical locations for research, travel allowances, and research group size. If the client is in the public sector, vehicles to make the modeling effort responsible to the public will be specified or neglected in this step. How high in the bureaucracy the modelers will be heard and on what terms they will be heard will also be determined.

More subtle trends are also established in this period. Do modelers and clients like each other? Is there to be mutual enthusiasm about the modeling project or mutual distrust? Will the modeler listen to the client? Will the client listen to the modeler?

Between the mundane and the subtle matters determined in establishing contact, the constraints that will shape the modeling study are fairly well defined. Where there is a client who knows the system well, the extent to which his knowledge will be incorporated into the model will be constrained by the communication channels established between modeler and client.

The size and complexity of the model will tend to be proportional to the funds spent on modelers' salaries. (There may be cases of modelers making a model simpler by investing more time in it, and even cases of larger staffs creating smaller models. The overwhelming tendency, however, is for complexity to increase in proportion to the time spent modeling.)

The quality of documentation will be heavily influenced by the time and resources allotted to secretarial, editorial, and printing costs and by the emphasis placed on documentation in early stages. Where there is both a user-client and a sponsor-client (for example, U.N.-sponsored models built for member nations), the relationship between the user-client and the modeler is bounded by the arrangements made to convey the model to its intended users.

Planning a model is no simpler than setting up a business enterprise, planning a battle, or designing a house—and should be approached with no less care. Planning a modeling effort involves coordination of limited resources toward an end. The resources are diverse and their functions complex; humans, machines, institutions, and money are all involved. A modeling study is more likely to succeed if its organizational and material underpinnings are carefully thought through. Investment in the wrong technology, forgetting to account for "human factors," lack of attention to user (consumer) needs, overly timid or overly ambitious plans—all can easily undermine a modeling effort.

Start from a vision of how you want the modeling effort to proceed through each step of modeling. Imagine the real-world details of each step, how much time will be required to do a good job, how modeler and client should communicate, and what things are likely to go wrong. Make provisions for things that should happen. Take precautions against likely difficulties.

When envisioning how the study should proceed, review the histories of your previous studies and examine the procedures followed by others. Try to avoid repeating mistakes. Pick up on things that appeared productive.

Noncongruent expectations are a rich source of tension and misunderstanding between modeler and client. It is worth checking for them during the planning stage and routing out any that are found. The formula for this precaution is simple: frank discussion, arbitration, and formalized consensus. Have the client relate in detail what he expects you to deliver and what he hopes the model will do. Tell him point by point all the ways you will require his cooperation to do your job, what data you count on him to supply, what sorts of support and cooperation you expect when it comes time to implement the model. Quite likely one or both of you will find the other's notions naive. Continue discussion and arbitration until your pictures for who, what, when, and how are identical—and pragmatic. Finally, to prevent false expectations from reestablishing themselves, put the agreement in writing.

In this and the next step (conceptualization), both client and modeler are apt to become impatient to "stop fooling around and get to work." Resist that valuation. These steps are the most critical and the most difficult parts of the modeling process. Do not be misguided by the absence of immediate tangible evidence of production. Haste is folly when it is unclear where to go. As a *minimum,* allow a month planning per year of work. To keep financial pressure from rushing this step, you may want to adopt the trial contract strategy that has proven useful at Pugh-Roberts (see Weil, chapter 13). Start out with a contract for a few months' work. Build a prototype demonstration model. See how things go in your relationship with the client. After the trial period you and the client will be in a much better position to ascertain whether or not you want to work with each other. Presuming you do, you will be in a better position to design the terms for further work.

12.4 Conceptualization

Conceptualization, like most processes, has tangible, observable aspects that can be studied to figure out how the process works. Having developed an operational model of the conceptualization process, we may devise means to control it and to move the process into more desirable modes of behavior.

Or such is the theory. Here we sketch a rough beginning towards devising an operational model of conceptualization by selecting three tangible aspects of the process—enthusiasm, organization, and synthesis—and broadly characterizing their behavioral trends. Enthusiasm refers to the level of energy at which the modeler pursues concepts, organization to the level of orderly information accrued in lab notebooks and other materials that support concept development, and synthesis to the extent to which information and concepts are brought together in a useful purpose-oriented manner.

The behavior of these aspects over the course of a modeling study seems to depend on the modeler's personality. I have observed three fairly stereotyped patterns, as shown below in figure 12.1. For the sake of pseudo-scientific puffery these have been labeled in Linnean fashion as *Conceptulus methodica*, *C. effusa* and *C. fructescens*.

C. methodica (common name: the drudge) works carefully assembling data, collecting accepted theories, and drawing diagrams. If he has whims, he doesn't follow them, nor does he venture too far from the obvious when drawing boundaries. His work is typically orderly but not inspired. That is, he does not create new theory. *C. methodica* thrives on well-cultivated theories. He is most comfortable readapting a generic structure, and is apt to become quite lost starting from scratch.

C. effusa (common name: the dreamer) is *C. methodica's* opposite. Judging from the frequency with which he uses the word "interesting," we might hypothesize that avoidance of boredom is his mainspring. He avoids the ordinary as assiduously as *C. methodica* clings to it. Typically,

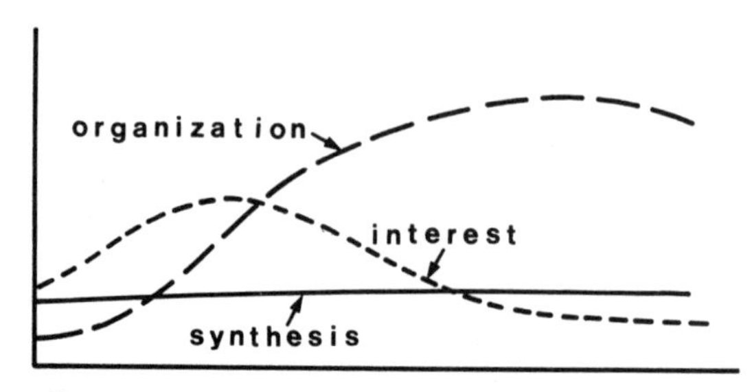

C. METHODICA

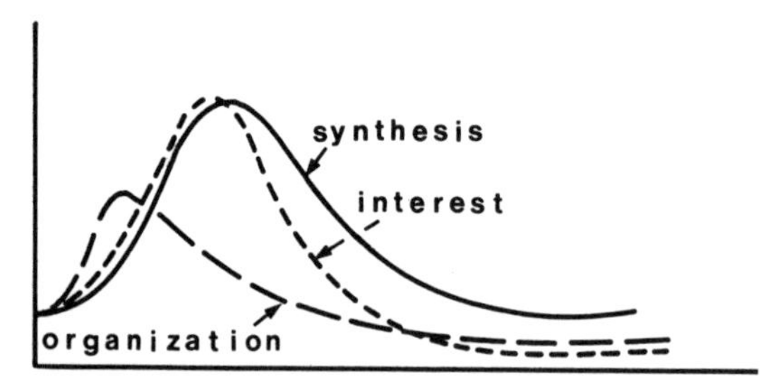

C. EFFUSA

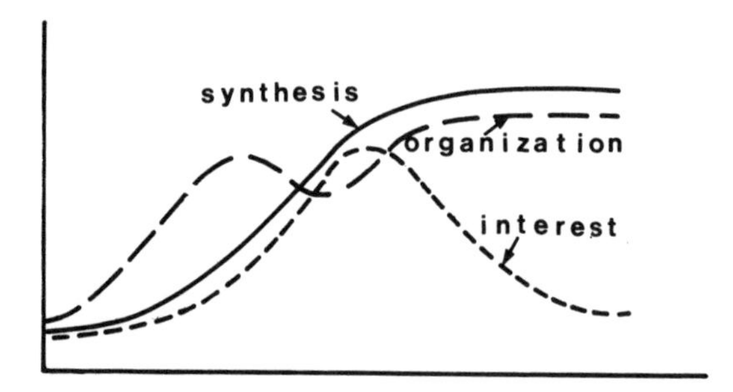

C. FRUTESCENS

Figure 12.1
Characteristic development of organization of material, interest in the problem, and synthesis of theory for three races of conceptualizers

he favors wide boundaries. Chasing whims and ethereal syntheses are his specialties—often to the exclusion of order. *C. effusa* does best in areas where there is no established theory. However, he is more likely to simply spin off new theories than to develop any solid theoretical structure.

C. fructescens (Common name: the fruitful) is generally considered the ideal. Though he may suffer temporary lapses in organization and may lose interest in a given piece of work, he—unlike *C. methodica* and *C. effusa*—pulls through at the end with both a high level of organization and a high degree of synthesis. The generic models *C. methodica* uses and *C. effusa* avoids are products of *C. fructescens* work. *C. fructescens* is clearly distinguishable from other races of conceptualizers in his ability to select the essential and reject the dross. *C. fructescens* neither follows superfluous whims, as *C. effusa* is prone to, nor works in excessive detail as *C. methodica* tends to. Rather he moves toward a clear, clean theory.

In the short term the modeler has little choice but to live with his innate conceptualization traits. Methodical people should concentrate on problems that are conceptually straightforward, such as adaptions of generic models and engineering-type structures that do not require generation of original social theory. Effusive conceptualizers should seek situations in need of theoretical ground breaking.

In the long term it would be more useful to be able to move both methodical and effusive styles toward the more desirable reference mode presented in the description of *C. fructescens*. Many people assert that conceptualization, like creativity, cannot be taught; there don't seem to be any cases on record to dispute that assertion.

There is, however, some evidence that conceptualization may be influenced by variables less nebulous than muses and innate genius. The time phasing of the process Jay Forrester describes himself as using is approximated in figure 12.2a.

For a new generic model, roughly two years is invested in sorting through information about the system, identifying recurrent behavioral trends associated with a problem, and isolating the state variables controlling the trends of interest. This done, the formulation of rate variables and construction of the initial model takes a number of weeks. Thereafter conceptualization phases into a period of extensive testing and observation of model behavior, which in turn leads to refinement of both the model and of the analyst's conceptual understanding of the system.

By contrast, the time phasing of a more normal modeling study (my observation) is approximated in figure 12.2b. The modeler more or less rushes into the study. Often he begins with a rough idea of what his main

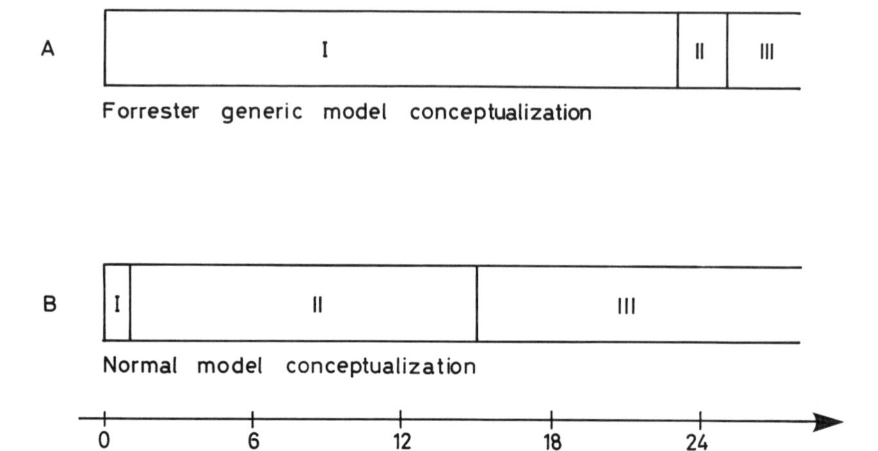

Figure 12.2
A comparison of time allocated to phases of model conceptualization: (I) identification of trends and state variables, (II) construction and reformulation, and (III) testing, interpretation, and later stages

state variables are to be. Within a month he is constructing the initial model. Construction is a relatively lengthy—and messy—process. Frequently it involves bouts of problem redefinition, expansion of model boundaries, and incorporation of new state variables.

Correlating the Forrester process with Forrester results and normal process with normal results does not imply that a long gestation period is the key to fruitful conceptualization. It does, however, offer the whisper of a suggestion that production of model concepts, like other forms of production, is not random: that there may be systematic relationships between what goes into the process, how it is organized, and what is produced. If such systematic relationships do exist—and we can develop an understanding of the system—it is likely that we can learn to control systematically the conceptualization process. If the development of such managerial know-how does not revolutionize conceptualization —if modelers admit to the conceptualization control scheme's validity and ignore it in practice—we will at least have developed grounds for perfect empathy with clients who balk when it comes to implementing structural change.

Warning! Once attained, conceptualizations tend to erode. Modelers become mesmerized by their models. They get to know each equation personally, and lose the ability to visualize the real-world situation the equations represent. Symptoms of conceptual erosion are often striking in model documentation. In extreme cases, conceptual understanding degenerates into a nitpicking detailed analysis without an overview. Documentation moves through the model line by line but loses sense of the whole structure. Mathematics predominates over social awareness. Form outvoices content. Great care is taken to justify everything, but nothing earthshaking is said.

A clear mental model must be maintained throughout the process of building a formal model. If the modeler ever ceases to sound like he has a clear understanding of the real world, something is disastrously awry. You simply cannot build a good model without a clear image of the part of the real world you are simulating. In the long run, a clear understanding of the real world is of far more value than the model itself, particularly to the client. A clear understanding must always be kept in the foreground. The greatest gift the modeler can give is a conceptual grip on the wheels of the world.

Modelers must discipline themselves to step back. They must look up from their models and take the real world more seriously than they take their representations of it—regularly throughout the modeling process. Along with book research they should observe the real world. They should talk with the people whose decisions are information streams in the model. Wherever possible, they should observe the plant and physical operations the model simulates. When they read, they should not confine their readings to economics and statistics but should read history, philosophy, anthropology, and novels as well. And they should feel that those readings are as pertinent to their work as are economic tracts.

Conceptualization must begin from realism. Nothing should be incorporated into a model until the modeler can imagine, in a tangible way, how it happens in the real world. Before introducing a money flow, the modeler should imagine the people who handle the money—the ways they are feeling, what they look like, the circumstances in which they live. Production functions should not be written until the modeler can visualize the buildings and the machines, the workers and the material with which they work. Population-growth functions should not be written without asking, "What do children mean to these people?"

A deliberate effort must be made to maintain realism. After translation into mathematical form, concrete understandings easily drift off into abstractions and technical mumbo-jumbo.

Second warning! Ongoing conceptualization frequently gives rise to deep, fundamental questions such as, "What does this model say about the way the real world operates? Is that how the world behaves? Does the model address important questions? Are the solutions this modeling study poses within the client's power to implement? Am I working for the right client? Am I building the right model?"

When sufficiently mild, such questions are highly beneficial: they furnish the jolt needed to initiate rounds of model refinement and improvement. However, the effect of questioning on a deeper level can be quite troublesome. Like the third-year medical student asking, "Do I really want to be a doctor?" Rethinking one's conceptual premises can lead to ulcers. There will almost always be some material about which the modeler can develop grave doubts. Yet pragmatic and institutional constraints commonly make it unreasonable to go back and rework the areas of uncertainty. If the model is a group effort, each member of the group may develop his own set of qualms about the model's conceptual base, and deep questioning may fragment the modeling effort. Moreover, modelers who have signed a contract and have deadlines to meet cannot easily backtrack because they have become unsure of their direction.

In modeling utopia, such questions would emerge during conceptualization to be openly discussed among modelers and model clients. The discussion would lead to a new, more useful problem definition. The modelers would then proceed without nagging doubts about what they were doing. The client would end up with a model that filled his needs and modelers who were sure enough of the merits of their work that they would be glad to take the pains to record it well and deliver it to the world with enthusiasm.

In practice, few models fit the scheme of modeling utopia. Too often the problem becomes redefined in the modeler's mind half way through the modeling process. By this time, the modelers are already too committed to their original problem definition to change the model. Thus they finish off the model according to the original problem definition and put their enthusiasm into looking for a new contract that will allow them to work on a new model with the knowledge gained from the last model. The old model, with which the modelers have become disenchanted, is completed and documented only sufficiently to meet contractual obligations. The commitment needed to communicate and implement a model is lost.

Both the need for and the destructiveness of hindsight questioning are apparent. Whether modeling can be managed to avoid the destructiveness remains to be seen. More time for the initial conceptualization—

closer modeler-client relationships or institutionalizing one or more periods to rethink the original problem definition, as advocated by experienced modelers at Pugh-Roberts (Weil, chapter 13; Roberts, 1972), might help to bring the initial model into a form so sound that grand questions will not prevent its satisfactory completion.

12.5 Construction

Models are designed to solve problems and are not an end in themselves. The kinds of models constructed are determined by the needs of the problem to be solved. (Anyone involved in model building can testify to the difficulty involved in being objective about this. It is *very* easy to let models become an end in themselves.)—T.J. Manetsch (1974)

Model construction consists of translating the conceptualized structure into a form that the computer can digest. To the modeler, model construction is home ground. He may never have studied how to conceptualize or how to relate to a client, but he has had years of training in model construction. Generally he is fond of the trade and enjoys using its tools.

Having a model to construct is having an elaborate puzzle to solve. The modeler proceeds with enthusiasm. He pulls out all his shoptalk and begins discussing with his fellow modelers the relative merits of one design feature or another. Talk turns to reference modes, slopes of table functions, dominant loops, oscillatory behavior, and modelers begin teasing each other about keeping the model simple while they wander off into bogs of complexity.

Tools cannot be depended on to remain passive. If you use system dynamics tools you will be coerced into representing your system as a higher order nonlinear system—whether that's the way you conceptualized it or not. Attempts to meld system dynamics with input-output or linear programming have high failure rates, and if you conceptualize a system that borrows from other schools of modeling you are apt to end up with a large mess on your hands. Make sure, well before you begin to construct a model, that the tools you intend to use are suited to the concepts you intend to formulate; don't count on having them bend to suit your needs.

Client and Construction

Of all the steps of modeling, there is none more alien to the client than construction and none where he is more likely to be left out. The buzz of shoptalk and jargon which arises as the modelers descend on their puzzle is nonsense to all but the initiated. Even if the client were educated in modeling, he would at this stage be hard put to keep abreast of what is going on. (Indeed, the modelers themselves may be lost a good part of the time.)

In short, while the model is translated into something the computer can understand, it is being translated into something the client cannot understand.

Some loss of client understanding during model construction is unavoidable. However, that loss should and can be minimized if

1. contact does not cease entirely during model construction,
2. modelers introduce the client to their jargon slowly and keep it to a minimum,
3. clients insist that modelers make what they are doing clear and inform the modeler when they cease to understand what is going on,
4. the model construction phase does not take so long that the client forgets that the model exists, and
5. the model becomes no more complicated than its assigned purpose requires.

The above "ifs" will not be met spontaneously. Their fulfillment is best assured if the difficulties during model construction are anticipated in the contact-establishment phase. A modeling study that starts out with a well-timed reporting schedule, a consensus to maintain communication, and a realistic attitude toward the difficulties of communication is far less likely to run aground during model construction. To the extent that communication does fail, despite advance preparation, the damage will be much less serious and much easier to repair.

12.6 Testing

Testing is the intellectual highpoint of the modeling process. In a sense, formal models are built to allow testing. Were mathematical models not amenable to a diverse spectrum of testing procedures, they would have little advantage over verbal models. Procedurally, too, testing is a climactic activity. Suspense tends to build from the time the model is conceptualized to the time it is ready to be tested, as the modeler wonders how it will work.

Sensitivity and policy testing are recognized as important (Forrester,

1961). How sensitivity and policy tests should be conducted is taught largely through the use of examples and through apprenticeship. In other words, testing form is a matter of judgment. Judgment is gained by experience.

We do not need to look very far to find poor judgment in model testing. Many modelers suffer from the inclination, when faced with a large model and an astronomical number of potential subjects for testing, to concentrate their efforts on "tuning" the model rather than testing it. Parameters are adjusted to attain a better historical fit more frequently than they are subjected to changes that might challenge the robustness of system behavior. Although real-world decision functions and information flows are frequently full of noise (Forrester, 1961), it is rare for modelers to test the sensitivity of their models to different amplitudes and types of noise. Extreme parameter combinations are seldom investigated, and structural changes are scarce. Thus model behavior under extreme conditions—the very realms in which nonlinearities become important and interesting results frequently occur—often goes unobserved. Even worse, little serious thinking seems to go into model testing. In almost two years of weekly seminars and daily lunchtime discussions within a modeling group, I have yet to hear anyone seriously discuss how to structure testing of his own model. Most people, myself included, appear to begin testing informally, as part of the debugging and refinement process. From informal tests they develop an intuitive grasp of how the model functions. From this intuitive grasp—which could be quite inaccurate (as we all know, intuition has a rough time with higher-order systems)—they go on to structure sensitivity tests. From the results of sensitivity tests and a notion of what policy tests will produce the desired effect on the client, they conduct policy tests. Whereupon testing is considered complete unless outside criticism intervenes.

Why does testing so frequently become superficial? For one thing, modelers often become so absorbed in revision and elaboration of structure that they leave no time for careful testing. This use of time is somewhat like preparing an elaborate meal and throwing it out uneaten. Simulation's main advantage over other forms of analysis is that it allows the modeler to see what happens when all assumptions operate simultaneously. If he quits after assembling his assumptions and doesn't take the time to observe the results of their interaction, in detail and under carefully devised experiments, he might as well be writing essays or drawing diagrams. For another thing, rigorous testing runs counter to the grain of intuition and habit. The verbal theories with which we were raised cannot be tested as simulation models can; they are too inexplicit to allow detailed examination of structure and behavior and too inflexible to allow experimentation. Unless a deliberate attempt is made to establish and maintain rigorous testing procedures, modelers easily revert to testing formal models with little more care than they would employ in evaluating a verbal model.

Moreover, there may be active subconscious resistance to testing. Few of us relish manipulating our models in ways that could invalidate either our conceptualized structure or our intuitive understanding of system behavior.

Such anti-testing forces should be ruthlessly opposed. Validity testing must not be allowed to degenerate into an attempt to demonstrate that the model is valid. It should be a serious attempt to locate places where the model, or the modeler's understanding of it, is *not* valid, leading to improvement and refinement of model structure and the modeler's structural understanding. The exercise should be undertaken with the same vicious skepticism one would use in test driving a used car—there is almost always something wrong under the hood. The point is to locate and correct the problems, not to paint over the rust spots.

Specifically, testing should involve careful observation of model variables under a variety of experimental conditions. The modeler should identify the structural reasons for variable behavior and for differences in variable behavior under different experimental conditions. Then he should question whether the model's structural causes are plausible in the real system. (If they aren't, it is time to think about revising the model structure.) To derive full benefit from tests, the modeler should be explicit about how he expects the model to perform in each test and watch carefully for ways in which model output deviates from expected behavior. Anomalies between expected and observed results are a signal that either the model or the modeler's understanding of model behavior is unrealistic. In either case, there is a lesson to be learned.

Given a set of hostile critics pressuring for tests that are largely irrelevant to the model's purpose and structure, the modeler will be driven toward defensive, rather than insight-seeking, tests. Thus the heat is on for tests that "prove" rather than "improve" models.

The modeler may avoid this destructive situation in a number of fashions. He can hire an outside critic whose judgment he respects. He can structure inside criticism into the modeling process by deliberately assigning members of the modeling term to criticize the work of other team members. Useful criticism seldom appears spontaneously. If the modeler wants it, he must actively solicit criticism and reward it when it comes.

12.7 Documentation

> . . . The writer either has a meaning and cannot express it, or he inadvertently says something else, or he is almost indifferent as to whether his words mean anything or not. This mixture of vagueness and sheer incompetence is the most marked characteristic of modern English prose. . . . As soon as certain topics are raised, the concrete melts into the abstract and no one seems able to think of turns of speech that are not hackneyed: prose consists less and less of *words* chosen for the sake of their meaning and more and more of *phrases* tacked together like the sections of a prefabricated henhouse.—George Orwell (1950)

Behind the question of documentation responsibility lies the larger question of the modeler's identity. Is the modeler a scientist? Is he a consultant? Is he an agent for change? Is he a servant of the quest for truth? Is he working for the client's interest? Or is he working for "the good"? All of the above?

If the modeler is a scientist, his documentation responsibilities are straightforward. By the general practice of the sciences he should (1) review the literature to establish the position of his study within the body of established knowledge, (2) state his problem or hypothesis, (3) describe his method in enough detail that an independent scientist could use the same procedures and attain the same results, (4) describe his results, (5) interpret his results and draw conclusions. When using computer models, he should also state his assumptions and describe data sources.

If the modeler is a consultant, his documentation responsibility is entirely pragmatic and highly circumstantial. He should document as best serves the client's interest. The client's interests depend on what he wants to do with the model and how well he grasps the model without documentation. If the client has been actively involved in the modeling process and understands the model, his needs may well be satisfied by a set of outlines and a few charts detailing the features of the model he most needs to know and is most likely to forget. If the model was completely ad hoc, if it answers a question and is of no further use once that question is answered, careful model documentation may be a luxury. On the other hand, if the model was intended as a tool for ongoing use in planning, client interest may require extensive documentation. User manuals will be needed for maintenance, updating, and operation, as will numerous explanatory works to assist in interpreting the model's output and communicating its results. If the client uses models frequently, he may benefit from such documentation as it will allow knowledge gained from one modeling study to be transferred to the next.

The modeler may have a message he wants to get across and the model may be a device to help him make his point. In this case his documentation is essentially propaganda and should be written to be convincing. It should be in short, undetailed, clear, and hard-hitting language. Methodological discussion should be minimal, and nontechnical. Main points only should be stressed.

Finally, the modeler may be an academic. Many modelers are professors living in a publish-or-perish world or graduate students writing dissertations. If the modeler is in one of these positions, he will want to document to meet the standards of the academic, but not necessarily scientific, community. His documentation form will be directed by the format of one or another academic journals or by a thesis committee.

One documentation job cannot serve all of the above functions. Scientific documentation is seldom what the client most needs for his own use. Activist documentation cannot easily meet scientific standards or the standards of the academic community. Should the modeler then document in five different ways to answer all responsibilities? I leave that question to you, but with a recommendation that the decision should be made deliberately in the step of establishing contact. If documentation responsibilities are not clearly established and the needs for documentation are not allowed for, documentation tends to become an afterthought. Afterthought documentation is seldom of good quality.

When a documentation style is selected, the following should be considered:

1. Documentation is a time-consuming process. The process of preparing a public report plus a scientific or academic report of a modeling study may double the time and expense spent on a modeling effort. If good documentation is desired, time and money should be allotted for it in the contract.
2. Documentation may improve model quality. Modelers are apt to be sloppy in one way or another. A lot of poor modeling is done and never detected. If required to present their work and make it accessible to criticism, modelers are more likely to catch their mistakes and less likely to use technically unjustifiable procedures.
3. Few modelers write well. Preparation of documentation for the general public, be it explanatory or persuasive, will probably require

editorial assistance and may require hiring a writer. If there is to be a report to the public, provision should be made to assure adequate writing skills in the contract agreement.

4. Modeling has great potential as a tool to mystify and confuse the nontechnical world. If models are used as tools in public decision making and are not documented in such a way that the public can understand them, they are, in spirit, technocratic and undemocratic.

5. Documentation is important for spreading model-generated knowledge. An undocumented model contributes more to the modeler's understanding of the system than it does to the world's understanding. If a model is well documented, the model-generated knowledge can be passed on to the technical community through scientific and/or academic documentation, to the public through plain language documents.

6. Computer modeling is a young discipline, and its field record is poor. If modeling is to improve and grow, knowledge gained in one study must be passed on to other studies. It is also important that models be subjected to criticism, both from the technical world and from the nontechnical world (particularly that portion of the nontechnical world acquainted with the real-world situation the modeler is simulating). Such criticism requires that models be given both technical and nontechnical documentation.

7. If a model isn't important or if its documentation is particularly obtuse, no one is likely to pay much attention to the documentation. Generic models and models with controversial subjects therefore have greater need of documentation than do case-specific models designed to answer questions no one worries about. Nontransparent documentation may not be worth printing.

8. Not much is known about model management or about the process of modeling. Documentation of managerial aspects might be fruitful for the development of management techniques.

12.8 Implementation

The first problem of modeling, which we have mentioned previously, is how to build the right model and interpret it accurately. Assuming this has been accomplished and potentially valuable conclusions have been drawn from the model, a new problem arises—how to put the conclusions to use. Modeling is not functional unless it leads to an improvement of operational conclusions *on the client's side*. The modeler's conclusions are absolutely useless unless they are transmitted to the client as operation conclusions.

Failure here is more common than success—in large part due to unrealistic expectations. In modelers' utopia, where most modelers seem to presume they are operating, a deeply distressed client comes to the modeler with a problem, crying, "What can I do to avoid this ruinous fate?" The modeler steps forth, performs an analysis, and offers a solution. The solution is accepted and used by the client. Of course, the solution works beautifully, and the client is eternally grateful.

By contrast, a cynic might describe the real-life situation as tending toward the following: modeler and some funding source come together and agree to build a model. The model purpose is designed as "furthering the understanding of *XYZ*" and "exploring the usefulness of the *PQ* methodology." The modeler receives funds and goes off to build his model. Two years later he brings back his masterpiece. The funding source cannot use it, but they know the chief of the BS division of the CDE agency which formulates *XYZ* policy. An appointment is secured. The division chief listens politely and looks over the model output, but does not call the modeler back or ask for further information. The modeler's interests move on; the model fades out of the scene entirely.

What happens in modeling utopia that fails to happen in real life? Essentially, in utopia the transfer of model-generated ideas is smooth and automatic, while in real life the gap between the modeler's and the client's ideas is so great that little or no transfer takes place.

Many things that the modeler takes for granted are invisible to the client. He is seldom given ample time to become familiar with the definitions of model variables before the modeler assails him with a detailed causal loop diagram and begins explaining structure-behavior relationships. If he tries to pull back to look at model assumptions—the natural thing for him to do at this point—the modeler is likely to cut off his inquiries with the observation that structure determines behavior. Eventually the client gives up.

Recommendations on how to avoid such problems have been made by Ed Roberts and others (see Roberts, 1972). Most of the strategies suggested focus on (1) involving the client in conceptualization, (2) closer more frequent modeler-client communication throughout the modeling process, (3) starting simple, and (4) including client assumptions in model structure. In general, such procedures offer a means for keeping the modeler's and the client's respective conclusion structures growing in synchrony throughout the steps of modeling.

An alternate strategy more appropriate to situations where close client contact is not possible (especially the situations where the client is the general public) would be to translate model-derived structural and operational conclusions completely into verbal form, and then let the translation compete with other mental models on mental model terms (see, for example, Randers, 1976, and Budzik and Meadows, 1975). At its best, the end product of such a strategy would be an essay, such as Garret Hardin's "Tragedy of the Commons" or Malthus's essay on population: a straightforward explanation of how the hypothetical system works, how it resembles the real world, and what factors influence the system's behavior.

System dynamics studies seldom yield conventional answers. Our paradigm gives us a propensity toward boatrocking, iconoclastic, radical conclusions. We tend to tell people that their previous actions have either had no impact at all or else aggravated the problem. We frequently insist that nothing short of drastic actions (that is, structural change) will make the system behave in the desired fashion. These traits tend to provoke emotional reactions.

That our conclusions are seldom neutral places an additional burden on the conclusion-transfer process. This burden would be eased if, before a modeling study is begun, the modeler and client would have a frank discussion about boatrocking. The modeler should make sure the client understands the propensities of system dynamics and is willing to expose his conceptual framework to major conceptual upheavals. If there is any doubt about client readiness for the modeler's type of conclusions, the modeler should seek a new client and/or the client should seek a new modeler. Thereafter the modeler should stay off his high horse and avoid posing as the problem-solver of modeling utopia. He must not presume his model will be accepted. If he aspires to see his model implemented, he must deal with real people, real institutions, and real inertia. He must remember that it is easy and safe for him to advocate change but that it will be difficult and risky for clients to implement it—in short, that there are usually well-based reasons for client resistance to model conclusions.

Note

1. The Korean Agricultural Sector Simulation Model was constructed under George Rossmiller at Michigan State University.

References

Budzik, Philip M. and Donella H. Meadows. 1975. "The Future of the Vermont Dairy Farm." DSD no. 50. System Dynamics Group, Dartmouth College, Hanover, N.H.

Forrester, Jay W. 1961. *Industrial Dynamics.* Cambridge, Mass.: MIT Press.

Hammond, John S., III. 1974. "Do's and Dont's of Computer Models for Planning." *Harvard Business Review,* vol. 52.

Kepler, Johannes. 1609. From *Astronomia Nova.* Cited in Arthur Koestler, *The Sleepwalkers.* Middlesex, England: Penguin Books, 1959.

Manetsch, T. J. 1974. "Basic Systems Theory and Concepts Underlying Construction of the Korean Simulation Model with Implications for Further Work." Department of Agricultural Economics, Michigan State University. East Lansing, Mich.

Mass, Nathaniel J., and Peter Senge. 1976. "Alternative Tests for the Selection of Model Variables." Alfred P. Sloan School of Management working paper no. 828–76, MIT, Cambridge, Mass.

Myrdal, Gunnar. 1968. "The Beam in Our Eyes." In *Asian Drama:* vol. 1, New York: Vintage Books.

Orwell, George. 1949. *1984.* New York: Harcourt Brace, Inc., appendix, p. 227.

Orwell, George. 1950. "Politics in the English Language." In *Shooting an Elephant.* New York: Harcourt Brace, Inc..

Randers, Jørgen. 1973. "Conceptualizing Dynamic Models of Social Systems and Lessons from a Study of Social Change." Ph.D. dissertation, Alfred P. Sloan School of Management, MIT, Cambridge, Mass.

Roberts, E. B. 1972. "Strategies for Effective Implementation of Complex Corporate Models." Cambridge, Mass.: Pugh-Roberts, Inc.

Senge, Peter M. 1975. "Testing Estimation Techniques for Social Models." System Dynamics Group working paper no. D–2199–4. MIT, Cambridge, Mass.

The Evolution of an Approach for Achieving Implemented Results from System Dynamics Projects

13

Henry Birdseye Weil

13.1 A Review of the Implementation Problem

Fifteen years ago, when I first began building system dynamics models, I thought very little about implementation. The approach was quite new then, and I was preoccupied with its compelling intellectual appeal and its technical facility. Now, it is fair to say, I am most concerned with the problem of achieving implemented results from system dynamics projects. Moreover, it has become a problem of critical importance to everyone in the field.

First of all, the sponsors of system dynamics projects deserve their money's worth, both as a matter of professional ethics and a basic tenet of good business. A private corporation that hires a consultant generally expects results of immediate value which can and will be implemented. Public policy analysis is admittedly a less clear-cut case, but here, too, projects which have a definite impact on people's thinking and actions are more valuable than those which don't. It is naive to think that sponsors are not aware of the implementation problem. On the contrary, they are increasingly demanding evidence of prior success in this regard.

In addition, work which leads to implemented results is far more satisfying. I can really speak only for myself and my close associates,[1] but I would be surprised if many others did not share this sentiment. We think of ourselves as *effective* professionals, because we can see the impact of our work. "Being effective" and "having impact" is an absolutely essential part of our concept of professionalism. It is depressing when, for whatever reasons, a project fails to produce meaningful implementation.

Furthermore, success at implementation is necessary to preserve and enhance the credibility of the system dynamics methodology. People all too easily reject an approach that does not seem to be producing results of immediate value. This tendency may be unfortunately shortsighted, but then shortsightedness is not a new trait of the human animal. In some quarters, system dynamics may already be stereotyped as only good for taking a very long-term view of very aggregate problems, and not capable of producing anything sufficiently detailed, specific, and practical to be usable in the short run. Of course, nothing could be farther from the truth. But the only persuasive way to answer such criticism is with an ever-growing number of clear examples to the contrary.

Despite its importance, meaningful implementation is an elusive goal. Success occurs only when all of the essential ingredients are present. First, the results of the project must, in fact, be implementable. This statement may sound obvious, but many projects offer recommendations which are technically "right," but at the same time too extreme, or too unconventional, or too inconsistent with established social/political structures to stand a realistic chance for being implemented. This shortcoming is an easy trap for a model builder to fall into. He is, after all, an outside technician who may well lack the perspective and sensitivity to determine properly what is implementable.

Furthermore, those who will have to take action must have a clear desire to implement. The sponsors need to understand, accept, and have confidence in the results before they will advocate implementation. They must attach a sufficient priority to the problems being addressed to justify the time and attention, the disruption, and the risks inherent in attempting to implement the results. In other words, there must be a significant client commitment.

And the environment must be properly receptive. Rarely are the sponsors the only people involved in implementation. In public policy analysis, the sponsors are often quite separate from the people who actually make and carry out policy. The policy-makers who have to be sold on project results are strongly influenced by their constituents and rivals, as well as by their advisors. If the political environment is not right, implementation will not take place, even though officials may agree privately that the recommendations are "theoretically correct." The same holds true in a private corporation. Management can accept your recommendations and direct that they be implemented, but resistance at lower levels or in parallel groups can easily defeat implementation.

Ed Roberts (1972) has outlined a number of factors which influenced these key ingredients for implementation success. This chapter is intended to document more fully lessons that my colleagues and I have learned about the achievement of implemented results from system dynamics projects. A series of case studies illustrates the evolution of our approach to implementation over the past ten years, and generalizes about the causes of success and failure.

13.2 Three Implementation Case Studies

Frohman (1970) has correctly remarked that "failure . . . is far more common in consulting than is generally recognized and is extremely embarrassing for both the consultant and the client. The mutual face-saving efforts that result often preclude using the experience as a learning situation for either the consultant or the client." A profession has to acquire a necessary maturity and confidence before it can be usefully self-critical. System dynamics has reached this point. My colleagues and I are not embarrassed to cite the mistakes we made in past years, because we have learned from them and are achieving implemented results much more consistently today.

How our approach to implementation has evolved can be illustrated through a series of three examples. All three were projects for private corporations, and all involve development of a system dynamics model. The first occurred in 1966, the second in 1971, and the third in 1975.

Case Study 1: Retail Food Chain

Background The client for this project was a large diversified corporation with activities in the supermarket, department store, food manufacturing, and restaurant fields. The project focused on several issues relating to the firm's retail food division and manufacturing division. Management considered the issues to be of significant mid-term importance, but, realistically, they could not be called "urgent."

The first issue was growing conflict between these two major divisions, which together accounted for sales of several $100 million. Each division increasingly blamed the other for its problems. However, no one could articulate a complete and correct picture of how in fact the two divisions affected one another's performance.

The second issue arose out of the first. The manufacturing division produced private-label food products that were sold almost exclusively in the company's own stores. These stores were operated by the retail food division. Goods were "bought" and "sold" between divisions at internal transfer prices, and the precise nature of this transfer pricing had an enormous impact on the accounted performance of both divisions.

Understandably then, in light of the general interdivisional friction, management was unable to agree on what constituted a proper transfer-pricing policy and what measures of divisional performance to use.

The third issue was related to the first two. Over the preceding five years, the company had invested very heavily in manufacturing facilities. A new bakery facility, which represented the largest capital investment ever made by the firm, was about to start operation. Manufacturing division profits had been lower than expected, and the top management was concerned that prevailing policies were somehow causing the division to underachieve.

The model developed during this project has been described by Roberts, Abrams, and Weil (1968). In brief, it centered around two related interdivisional flows: the flow of orders and goods, and the flow of cash. The determinants of retail sales were represented in considerable detail. The model was, for the most part, conceptually straightforward and easy to understand. The only conceptually advanced part of the formulation had to do with the process of managerial control with respect to multiple conflicting performance criteria. A very small model by our current standards, it contained 110 significant variables.

Outline of the Project Process The project got off to a fast start. Very little selling was required, nor did we have to prepare a formal written proposal. We had been approached by an MIT alumnus who was well acquainted with system dynamics and already convinced that it was the proper way to approach his company's problems. He occupied a key staff position in the company (assistant to the treasurer, who was the son of the firm's president and founder) and championed our involvement as consultants.

The first step was a series of discussions with top management regarding their perception of the problems. Next we conducted an initial round of "scouting" interviews at many levels in both divisions. The purpose of these interviews was to learn the organization and come to our own conclusions about the problems.

We then adjourned to our offices for about six weeks. During that period, the model was conceptualized, equations were written, and preliminary parameter estimates were developed from company-supplied data and our interview results. We subsequently had a series of lengthy meetings at which the model structure and initial simulation runs were discussed with divisional and top management.

Following these meetings, we again worked independently for a considerable time. We refined the model based on further interviews and data gathering. The model was initialized with year-end 1961 conditions, and a five-year historical simulation was produced for the period 1962 to 1966. Management reviewed this simulation and pronounced it "reasonable." However, we did not engage in any formal quantitative validation of the model. At this stage of the project, the model was accepted by company management as sufficiently realistic for policy testing.

The final phase involved the analysis of alternative policies. We examined policies in the areas of transfer pricing, divisional performance measurement, goal setting (for production, sales, and profitability), and managerial control. We presented our findings at a meeting with divisional and top management and submitted a reasonably brief (30–40 pages) nontechnical report.

Very few of our specific policy recommendations were implemented. General agreement did exist among the president, treasurer, vice-president of retailing, and vice president of manufacturing that the feedback concepts embodied in the model reflected the essential characteristics of their organization. They attempted to use the new conceptual framework provided by the model as an aid in thinking about the company's problems. Therefore the principal effect of the project was "conciousness raising." At the time, this outcome seemed perfectly satisfactory. In fact, there was a tendency, then, to think that "greater understanding" was the primary benefit to be expected from a system dynamics project. But, by our current standards, we were not very successful in achieving implementation.

Comments This project was rather typical of our approach ten years ago. Our principal point of contact in the client organization was not a line manager, but, rather, a sophisticated staff man who was favorably disposed toward system dynamics, understood what we are doing, and readily accepted the emerging analytical results. He was a very "comfortable" person to work with for these reasons. Communication was easy, and he posed few cross-cultural problems; he was sharp, quantitatively oriented, open-minded, rational—just like us! Unfortunately, he was not the president, nor the president's son, nor an influential member of line management. For all practical purposes, he was another outsider. The official sponsor of our project was the treasurer, but we had only intermittent contact with him.

Furthermore, the model was built and exercised almost exclusively by us, with little client involvement. At various points during the project, meetings were conducted with key company executives. These sessions covered the model's structure, the simulation results, and our policy recommendations. In each case, the managers were being asked to react to a reasonably finished product, not to participate in developing it. Our role in this project was that of systems analyst and researcher; people in the client organization served as information sources and as an audience for our work.

With the benefit of hindsight, clearly we were probing very sensitive areas. The question of transfer-pricing policy involved very significant financial stakes for the division affected. The question of proper performance measures involved very strongly entrenched traditions in the retailing industry. The questions of goal setting and control involved the personal styles and values of the president and other key managers. Our recommendations in these areas were viewed as "unconventional" and "controversial." We had a credibility problem. Although management generally acepted the structure of the model, they were not yet ready to accept its analytical implications.

Nor did the end product of this project allow management to come to terms with the results on their own, later. As noted previously, our only tangible end product was a nontechnical report. We did not deliver the model itself, with the detailed technical documentation required to examine it in depth, test it, and recreate our results. And we did not spend time developing the in-house capability to engage in continuing work with the model.

To summarize the salient characteristics of our approach ten years ago,

1. We tended to work independently as system analysts and researchers, with relatively little client involvement.
2. The client's role tended to be that of information source and audience.
3. Our closest contacts in the client organization tended to be with sympathetic and sophisticated staff people, rather than line managers.
4. We tended to build small aggregated models for the purposes of understanding system behavior and testing potential policy changes.
5. We tended to ignore questions of formal model validation and be satisfied with a model whose historical behavior was qualitatively reasonable.
6. The principal end product of our work tended to be a report.

Case Study 2: High-Technology Manufacturer

Background The second project was conducted for one of the world's leading manufacturers of information-processing equipment. Our specific client was the senior vice-president for research and engineering. He had requested our assistance in three areas which, he felt, were important to the success of his organization over the subsequent five to ten years.

First of all, he wanted to develop a framework that would help him communicate more effectively with the president and other senior executives. He felt that they did not understand the R & D process. He wanted to be able to show them how his division worked and where his problems came from. He was particularly anxious that they recognize that many problems in the R & D area were caused by people elsewhere in the company.

Second, he wanted to establish within his organization an ongoing strategic analysis process. He had come to his position from outside the company, and was not satisfied with the calibre of analysis he found. He was eager to address resource-allocation and technology-policy issues in a more sophisticated manner.

Third, he wanted to deal with a problem of erratic workflow through the R & D organization. This "workflow bunching" phenomenon manifested itself in several ways. The whole division went through cycles of overload followed by slack; these cycles were far more severe for individual R & D sections. Furthermore, the R & D process consisted of a sequence of phases, which could be viewed as a "pipeline" of sorts. Workload tended to surge down this R & D pipeline in a series of waves. Consequently, the organization often found itself understaffed in one area and overstaffed in another. Differences in skill requirements could easily produce simultaneous hiring and layoffs.

The model that emerged from this project was described by Weil, Bergan, and Roberts (1973). It represented the flow of work through multiple phases of R & D, starting with basic research and ending with major new products ready for large-scale manufacturing. The acquisition and allocation of human resources (scientists and engineers) was modeled in significant detail. Another important sector of the model represented the performance measurement and control process: comparison of actual performance with targets, estimation of future resource requirements, and revisions of performance targets.

The model was both large and conceptually subtle. Disaggregation with respect to R & D phase, characteristics of the workflow, types of resources, resource-allocation possibilities, and dimensions of managerial control produced a model with approximately 600 significant variables. The considerable size and complexity of this model was a direct result of an expressed client desire for "realism." Two aspects of the formulation were particularly subtle. First, the actual R & D workflow consisted of a relatively small number of large programs. Therefore our continuous representation of the flow was a substantial abstraction. Second, it was quite important to model various technical characteristics of the R & D work flow (for example, how well mastered was the technology being embodied in new products; how technically advanced were the products in R & D). The technological variables in the model also represented substantial abstractions. These abstractions make the model more difficult to understand.

Outline of the Project Process Unlike the project described in the first case study, this project started slowly. We had an initial two-day session with the vice-president and his key staff assistants to discuss the focus of the effort. Based on those discussions, we requested (and received) a small budget to pay for an initial conceptual definition of the system to be modeled and a detailed written proposal describing how we would proceed. These goals were achieved over a two-month period. During this definitional phase, the company made no commitment with respect to continuing the work. That commitment came with the subsequent acceptance of our proposal. We had met several times during the definition phase with the vice-president's staff. The proposal was a collectively accepted statement of project objectives, focus, process, and end products.

Once the project has been approved, the next step was selecting a task force. The project task force consisted of key individuals from the client organization: members of the vice-president's staff and managers of several of the most important components of the R & D division. In theory, the vice-president was chairman of the task force, but he and the other line managers attended only the major review meetings. The function of the task force was to work with us on developing the model, interpreting the results of the analyses, and formulating policy recommendations.

Task force members participated with us in an extensive series of interviews, both inside the R & D division and elsewhere in the company. We refined the conceptualization, presented it to the task force, and (based on their critique) refined it further. We then wrote equations and assembled that necessary data to parameterize the model.

The initial simulation runs were reviewed with the task force. The historical behavior of the model was examined and deemed "reasonable." No particular effort was made to quantitively validate the model. We jointly planned refinements in the model and established priorities in that regard. While performing policy analysis simulations, we had frequent meetings with the task force to discuss the findings. We produced recommendations with respect to resource-allocation policies and R & D program planning.

At the end of the project, we prepared *very* complete documentation, consisting of: (1) a management summary; (2) a lengthy discussion of the model's conceptual structure and the simulation results; and (3) detailed technical documentation, which included a write-up of every equation and parameter in the model, all of the important simulation output produced, and instructions for using the model. Furthermore, we "installed" the model on the client's in-house time-sharing system.

No immediate action was taken on any of our specific recommendations. Our conclusions were viewed as something *they* should study further—in-house, privately. This position was tenable because the model and the capability to use it had been internalized by our client. A senior staff analyst was assigned to work the model, and he did so for several years. He called upon us for technical assistance from time to time, but we never learned precisely what he was doing. In this case, we achieved a great deal more than just increasing client understanding of a problem. We created a policy analysis capability that continued to be used. By our current standards, this project was a partial success in terms of implementation.

Comments This project illustrates the significant evolution of our approach between the mid-1960s and the early 1970s. We had by then learned the importance of directly enlisting senior line managers (who generally hold the key to implementation) in our work. They tend to be harder to sell, more skeptical, less analytically oriented, more frustrating, and less available for work sessions than our in-house counterparts. But none of that matters. Their participation injects important perspectives and considerations that are not available from other sources, and allows them to acquire the understanding and confidence they need to act on our recommendations.

By then, we were consistently employing a project task force as the vehicle for securing client involvement in our work. We expected such a task force to be much more than a review board. We expected the members to *work with us* in information gathering, model conceptualization, interpretation of simulation results, and formulation of policy recommendations. Of course, this objective is easier said than done. One problem in the project discussed here was that the task force subdivided into two groups. The core group with which we worked quite closely consisted of three members of the vice-president's staff. We met with the full task force less frequently. The line managers, themselves, were not sufficiently involved to obtain the requisite understanding of our work. As previously mentioned, the model, while large, contained some significant abstractions that were not easily grasped.

We had also learned the importance of sufficient preliminaries before a project is launched. As Frohman (1970) points out, the early phases of a project are most critical in determining eventual success or failure in implementation. A formal definitional phase (as took place during this project) allows both the consultant and client to "size one another up." This initial phase is the consultant's principal opportunity to influence the client's view of, and expectations for, the project. The basic "tone" or "climate" of the project is established here. Decisions regarding scope and involvement undertaken at this point generally determine whether there will be a sufficient power base to achieve implementation.

Furthermore, our concept of the end products of a system dynamics project had evolved substantially. There was much more emphasis on the transfer of know-how, on creating and internalizing in the client organization an ongoing capability for analysis. Therefore we delivered mountains of documentation; we delivered the model itself; and we trained some client personnel to use the model.

To summarize how our approach had evolved by the early 1970s,

1. We actively sought significant client participation in our work. We employed a project task force as a vehicle for securing client involvement.
2. The client's role tended to be that of information source, information collector, critic, and trainee.
3. We were particularly anxious to work closely with senior line managers.
4. We tended to build large, complex models that clients would accept as "realistic."
5. We still tended to ignore questions of formal model validation, and to be satisfied with a model whose behavior was qualitatively reasonable.
6. We viewed the end product of our work as a transfer of know-how and capability.

Case Study 3: Insurance Company

Background The work described in the third case study was performed for a major diversified financial institution. As part of its aggressive growth and diversificaiton strategy, this company had invested over $100 million in a very promising new business area. At the time we became involved management was increasingly concerned with a series of "growing pains" that the business area was experiencing. They were anxious to review and, if necessary, revise the strategy being followed in that business area. As our work progressed, a clearer perception of the magnitude of the problems and changed circumstances elsewhere in the company heightened the urgency of the situation.

The project focused on several key strategic issues. First of all, management was uncertain of the ramifications of different rates of sales growth in the business area. Among the important considerations were: sales growth/profitability trade-offs, maintenance of market position and "momentum," adequacy of the "quality" (that is, longer-term profit potential) of sales, dilution of organizational capabilities, requirements for additional capitalization, and vulnerability to adverse economic and/or competitive conditions.

Second, the most effective set of policies for achieving the desired near-term rate of sales growth was unclear. A balanced set of policies had to be defined with respect to product mix, pricing, sales force compensation, sales force size, customer service and underwriting.[2] The impacts of changing economic, regulatory, and competitive conditions also had to be properly factored in.

Third, we had to assess the short-run/long-run trade-offs inherent in each strategic option. The most appealing strategies in the near-term might have significant adverse consequences in later years.

A very large and complex model emerged. Separate major sectors represented: sales force size, skill, and effort allocation; determinants (both internal and external) of sales force effectiveness; quality of customer service; performance measurement and managerial control; and accounting relationships. The sales force was disaggregated by skill level and type of organizational affiliation. The determinants of sales were modeled in considerable detail. Managerial concerns, priorities, and reactions were richly represented. The accounting sector was highly complex, disaggregating booked business into five age categories and calculating profits on both a cash and an accrual basis. In total, the model contained over 1250 significant variables. It has been described in detail by Weil, Pugh, Wright, and Veit (1974).

Outline of the Project Process Our client for this project was the vice-president in charge of operating the business area. He proved to be an extraordinarily astute and motivated client. Prior to our first meeting to discuss the possibility of a project, he prepared a paper outlining the issues and companion causal loop diagram. He assembled for this meeting a group of key managers who, he felt, might be involved if a project went forward. We talked about the issues, the system dynamics methodology, other projects we had conducted (with particular emphasis on implementation success!), the type of model we might develop for them, and how we would proceed. These discussions continued in several meetings over a four-month period before the project was formally launched.

Our project task force consisted primarily of people who had participated in the earlier meetings. Therefore, they had been actively involved in defining the project and deciding it was worthwhile. Furthermore, the vice-president maintained a very high level of personal involvement. He led the task force and immersed himself in the project down to the smallest technical details.

We began the project by interviewing each task force member to gain general background information. We then devoted several task force meetings to familiarization with the techniques that would be used in the project: basic concepts of system dynamics (for example, feedback rates and levels), the DYNAMO language, and computer time-sharing. We presented a series of relevant example models, reviewing how they were developed, their structures, and the ways in which they could be used. Simulation experiments were performed on a terminal in the meeting room. These sessions prepared the task force to participate significantly in model conceptualization.

To begin with, one full-day task force meeting was allocated to discussing the project's problem focus and how this focus translated into a necessary set of model boundaries. Then we developed a gross conceptualization of the entire model, which was thoroughly critiqued by the task force. Next we developed the detailed conceptual design of several "core sectors" of the model. This design was critiqued by the task force. The task force also provided extensive comments on parameter values.

Equations were written for the "core sectors," and initial simulations were performed. At this point, several members of the task force took the time to scrutinize the equations and simulation results. They wanted to satisfy themselves that the model was reasonable on a detailed level and to understand "where the simulation results came from." These people served as a continuing technical working group. Of great importance, though perhaps surprising, the vice-president and his comptroller were part of this group.

We then proceeded with an iterative expansion of the core model. One after another, new sectors were conceptualized, critiqued by the task force, implemented in DYNAMO, and added to the model. Each time the new equations and simulation results from the expanded model were reviewed with the technical working group. This process of model expansion occurred over approximately three months. Members of the technical working group spent about six hours per week reviewing and discussing the work during that period.

The model was substantially complete within six months of the beginning of the project, and was installed on the client's computer system. We intensively tutored several members of the vice-president's staff in the use of the model. They undertook model testing and some refinements in parallel with our efforts.

We devoted a considerable amount of time and attention to improving the historical accuracy of the model. Simulated values for a large number of variables were explicitly compared with historical data for the period 1970 to 1974. The model generally produced results within $\pm$ 10 percent of historical values; in some areas, the accuracy was consistently within $\pm$ 5 percent. We achieved a broad consensus that the base simulation was historically valid and the best existing estimate of what the future held in store. The client had assumed "ownership" of the model (in the psychological sense).

Our policy analyses took place in two separate phases, about six months apart. The first phase focused primarily on questions of growth strategy: sales growth/profitability trade-offs, investments required to maintain a strong market position in the future, changes in the management control structure to make growth a more orderly process, impact of economic conditions on sales growth. The general conceptual framework provided by the model, the initial analysis results, and our best forecast for 1975 through 1980 became inputs to management's determination of near-term growth targets.

The major company decision was to dramatically slow down sales growth in order to improve profitability. The question then became: What is the best set of policies for achieving this goal? The model was expanded in several sectors where more detailed answers were required. Policies with respect to product mix, pricing, sales force compensation, sales force size, customer-service expenditures, and underwriting were analyzed with the model. The results of these analyses significantly influenced key managers' perceptions of the issues and the policy decisions that were ultimately made. We consider this project an implementation success.

Comments This project typifies our current approach. We now recognize the importance of the project task force playing an active role in defining the focus and scope of work to be done. This role requires involvement at a very early stage, while the project is being sold. We regularly request prospective clients to include in preliminary meetings people who would probably participate in any project that might be started. We also regularly request prospective clients to prepare for such meetings papers that define the problems they are concerned about (including causal diagrams). We now believe that a client must participate at a fairly detailed technical level in the process of model development. This technical role means, first of all, more emphasis on familiarizing people with the techniques that will be employed. We encourage task force members to read portions of the system dynamics literature, we devote task force meeting time to the methodology, we engage in extensive "on-the-job" tutoring, and we often ask clients to send people to system dynamics courses.

We have also modified our approach to model building to make it easier for clients to keep up with us on a detailed technical level. Now we tend to develop a gross overall conceptualization of the model first, then produce the detailed conceptual design and the DYNAMO equations in several blocks. This approach is not so initially overwhelming, and it produces simulation results earlier in the project. As the model grows more complex, the client grows more sophisticated.

We now routinely build very large models, for several reasons. First of all, clients are more comfortable with and confident in a model that they consider "realistic." Since these attitudes are an absolute prerequisite for successful implementation, we are generally very accommodating to client desires for more detail. Second, both we and our clients are far more confident in models of demonstrable historical validity. We now routinely engage in extensive comparison of simulation results with historical time series and expect a model to be accurate within ± 10 percent. To achieve this degree of historical accuracy (except in trivial situations, such as constant exponential growth) requires a very elaborate causal structure. Our experiences with many models over the past four years indicate that high historical accuracy is a technically feasible goal. Complex behavior patterns can be accurately reproduced with a sufficiently rich and explicit causal structure. Recently, we have been using the "full information maximum likelihood" statistical techniques for model estimation described by Peterson (1976).

Furthermore, the detailed implementable recommendations we seek generally necessitate quite disaggregated models. Therefore we often end up representing multiple product lines, multiple phases of effort, multiple classes of resources, multiple market segments, and/or multiple dimensions of management control in our models. Finally, we have learned that working on high-priority immediate problems greatly increases the likelihood of implementation success. But this kind of problem focus often increases the need for short-term (one to five years) predictive accuracy, which, in turn, mandates a more elaborate model. The project described in this case study is a good example. An important aspect of policy analysis was determining the impact of various potential actions on 1976 financial results. We needed to judge accurately how strong the actions should be to attain the exact financial results sought by management. Management's confidence in the model was significantly enhanced when the model predictions for 1975 (made in January of that year) turned out to be *very* accurate.

This project also illustrated a further evolution in the end-product we try to deliver. We now expend more effort creating an in-house capability. This effort is a natural by-product of the greater technical involvement we now demand of our clients. In addition, we have learned that we cannot just deliver our recommendations and ride off into the sunset, in what Frohman (1970) cites as the "Lone Ranger syndrome." The process of considering and acting upon consultant recommendations takes time. It often necessitates additional analysis to answer unanticipated questions and lengthy meetings to talk through the implications of the various alternatives. Therefore continuing strategic consultation is generally quite important in the period after the final report has been submitted.

To summarize the further evolution of our approach,

1. We now seek the active involvement of the project task force in defining the focus and scope of the work.
2. We now consider it necessary for the client to be a technical contributor to model development and a model user.
3. We now routinely engage in extensive comparison of simulation results with historical time series, applying demanding standards of accuracy.
4. We now regard continuing strategic consultation as necessary to assist a client in considering and acting upon our recommendations.

The project described in this case study is indicative of the time and effort required to follow our current approach to implementation. As noted above, the project proceeded in two phases. The first phase, during which the original model was developed, policy analyses were performed, and the work was brought "in-house" by our client and was completed in ten months. The second phase of model expansion and analysis took four months. A combined total of about one man-year of effort was expended by me and several associates to accomplish both phases of work.

13.3 Conclusions

From the preceding case studies, our approach has clearly evolved in a significant fashion over the last ten years. As a consequence of these changes, we are consistently more effective in achieving implementation. We still have a lot to learn, but we have made substantial progress in the right direction. Let us review the highlights of the evolution that has occurred.

Client Involvement

Ten years ago we tended to work independently as system analysts and researchers, with relatively little client involvement. A client typically served as a source of information inputs and as an audience for presentation of our work. Furthermore, as Ed Roberts (1972) pointed out, the "client" is not an organization but rather an individual. Our closest contacts in those days tended to be sympathetic and sophisticated staff people, not line managers. Consequently, the people who were in effect our clients often lacked both the perspectives to make our work "real" and the authority to act on our recommendations.

Today, we actively seek significant client participation in our work. We try to function not as researchers but as strategic counselors and change agents. We employ a project task force as a vehicle for securing client involvement. We expect this task force to participate significantly in defining the focus and scope of the project, developing the model, and formulating policy recommendations. We now expect a client to participate at a fairly detailed technical level in the process of model development and use. We are particularly anxious to work closely with senior line managers; we have learned that they are the right people to have as clients if you want to have an impact.

The Process of Model Development

Our approach to model building has changed to facilitate greater client involvement and also to reflect the different kind of model we produce today. We recognize that it is difficult for a client to keep up with us on a detailed technical level. The larger the model, the worse this problem becomes. Furthermore, we recognize the importance of having preliminary simulation results as early in a project as possible. Producing some early indication of the ultimate payoff is a very important step in building client confidence. In their terms, simulation results are generally viewed as the "first tangible thing coming out of the project."

We now tend to develop a gross overall conceptualization of the model first, then produce the detailed conceptual design and the DYNAMO equations in several blocks. This approach is less overwhelming than our earlier practice of building the whole model at once. Now the client's sophistication and the model's complexity tend to grow together.

Until recently, we tended to ignore questions of formal validation and be satisfied with a model whose historical behavior was qualitatively correct. We now feel that our clients are far more confident in models of demonstrable historical validity, and therefore more likely to act on recommendations resulting from them. We consider model validity to be a key implementation issue. We now routinely devote a considerable effort to achieving high historical accuracy with our models.

Nature of the Models that are Developed

Ten years ago we tended to build small aggregated models. They were the minimum size required to understand system behavior and test potential policy changes. This practice was the conventional wisdom in our field in those days.

We now routinely build rather large models. By "large" I mean models with 1,000 to 2,000 significant variables. We have found that models of this size are required to satisfy client standards of "realism" (again, very important in establishing the comfort and confidence prerequisite for implementation), to achieve the desired degree of historical and near-term predictive accuracy, and to produce findings of sufficient detail that they are implementable.

We are increasingly confident in our models as forecasting tools. I now believe that is unnecessary and counterproductive to make excuses for our methodology with statements such as, "System dynamics models are not developed for forecasting; they are tools for understanding problems." Our models can serve both purposes. We feel that our clients have greater confidence in us because we are confident in our approach. We have found that *nothing* enhances our credibility more than correct short-term predictions.

The End Product of a Project

A dramatic change has taken place in what we consider to be the appropriate end products of our work. Ten years ago the principal end product tended to be a report. This is no longer true. We now emphasize the transfer of know-how and the creation within a client organization of an ongoing analytical capability. As a result we usually deliver extensive technical documentation, install the model on a client's computer system, and engage in considerable training of client personnel.

We have also learned that we cannot just submit our recommendations, move on to other projects, and expect implementation to occur. The process of considering and acting upon consultant recommendations often gives rise to additional analytical requirements. Clients often want to talk through the implications of the various alternatives open to them, perhaps to probe more deeply or to use us as a "sounding board." Continuing strategic consultation is an important factor in achieving implementation success.

Summary

Our current approach to implementation is eminently practical. It reflects what we have learned about the realities of using system dynamics models to help people structure and solve difficult problems. It recognizes the critical psychological and sociological aspects of bringing outside technical consultants into an organization. An especially important aspect of being practical is the ability to address problems of importance and urgency in a relatively short period of time at a cost that clients find reasonable. Our projects generally take six to twelve months to complete and involve anywhere from one-half to one and one-half man-years of effort. It would be a mistake to assume that our approach leads to extravagantly expensive and lengthy projects. Indeed, such projects could rarely be successful by our terms.

Although the case studies in this chapter describe models developed for private corporations, we have found that the same general conclusions about implementation apply to policy analysis projects for government agencies. Here too it is extremely important to build confidence in and commitment to the results. To get very far, one must create a critical mass of motivated people to "champion" a new point of view.

Table 13.1
Comparative summary of three projects

Case study	Sharpness of problem focus	Urgency of the problem	Position of principal client contact	Degree of client involvement	Nature of client involvement	Size of model	Validity of model	End products	Implementation success
Retail food chain	High	Medium	Junior staff person	Low	Information source	Small	Subjective	Report	Low
High-technology manufacturer	Low	Medium	Senior staff person	Medium	Information source, information collector, critic, trainee	Large	Subjective	Reports, model, trained people, counseling	High
Insurance company	High	High	Senior line manager	High	Information source, information collector, technical contributor, trainee, model user	Very large	± 10 percent of historical values	Reports, model, partially trained person	Medium

Some of the factors that are most critical in achieving successful implementation are summarized in table 13.1. They include

1. the sharpness of the project's problem focus,
2. the urgency of the problem addressed,
3. the organizational position of the client,
4. the degree of client involvement,
5. the nature of client involvement,
6. the size of the model developed,
7. the demonstrated validity of the model,
8. the nature of the project's end products.

As discussed by Roberts (1972) and demonstrated in the case studies presented here, each of these factors contributes importantly to the production of implementable recommendations, to the development of a clear desire to implement on the part of those who have to take action, and to the existence of a properly receptive environment in which implementation can occur. Without these essential ingredients, it is not possible to achieve implemented results from system dynamics projects.

Notes

1. The author is vice-president and director of Pugh-Roberts Associates, Inc., Cambridge, Massachusetts. For the past sixteen years, his firm has specialized in applying system dynamics to many complex questions of corporate strategy, regulatory policy, economic planning, and the design of government programs. The author and his colleagues have engaged in over two hundred projects of this type.

2. The term "underwriting" refers to the practice in the insurance industry of screening potential sales to evaluate the risks involved and to determine the conditions under which the company would accept the business.

References

Frohman, A. L., and D. A. Kolb. 1970. "An Organization Development Approach to Consulting," *Sloan Management Review,* vol. 12 (Fall), pp. 51–65.

Peterson, D. W. 1976. "Statistical Tools for System Dynamics." *The System Dynamics Method.* Edited by J. Randers and L. Ervik. The Proceedings of the 1976 International Conference on Systems Dynamics, Geilo, Norway. Copies available from Resource Policy Group, Forskningsveien 1, Olso 3, Norway.

Roberts, E. B., D. I. Abrams, and H. B. Weil. 1968. "A Systems Study of Policy Formulation in a Vertically Integrated Firm." *Management Science,* vol. 14, no. 12, pp. B-674–B-694. Also in *Managerial Applications of System Dynamics.* Edited by E. B. Roberts. Cambridge, Mass.: MIT Press, 1978.

Roberts, E. B. 1972. "Strategies for Effective Implementation of Complex Corporate Models." In *Managerial Applications of System Dynamics.* Edited by E. B. Roberts. Cambridge, Mass.: MIT Press, 1978.

Weil, H. B., T. A. Bergan, and E. B. Roberts. 1973. "The Dynamics of R & D Strategy." Proceedings of the Summer Computer Simulation Conference, Montvale, N.J.: AFIPS Press. Also in *Managerial Applications of System Dynamics.* Edited by E. B. Roberts, Cambridge, Mass: MIT Press, 1978.

Weil, H. B., A. L. Pugh, R. D. Wright, and K. P. Veit, 1974. "Growth Strategy In a New Business Area: A Simulation Analysis." Presented at the Summer Computer Simulation Conference. Copies available from Pugh-Roberts Associates, Inc., 5 Lee Street, Cambridge, Mass. 02139.

A Modeling Procedure for Public Policy

14

Lennart Stenberg

14.1 Introduction

The field of application of system dynamics has changed. The technique was initially developed to aid in corporate policy making. Since then, applications have been extended to various areas of public policy where most of the work is now underway. This means that the system dynamics methodology is used today to deal with problems other than those for which it was developed. That change in problem focus will have to be accompanied by an evolution in methodology.

The situation of analysts working out unrealistic solutions to irrelevant problems is all too common. The original conception of system dynamics tried to assure that analysis and implementation remained integrated into one process. The modeling tools were developed with the clear intention of facilitating the interplay between managers' mental models and the analysts' formal models. Making the modeling work transparent to the managers reduces the risk of the analyst losing contact with reality and makes it possible for managers to effectively contribute to and learn from the modeling process as it evolves. As long as managers remain in close contact with the model-building exercise, the issue of model validity changes from the abstract question of how true the model is to a question of whether the model helps the manager understand his situation better.

Unfortunately, there are few detailed accounts of how system dynamics modeling has been integrated in the policy-making process of a company or an organization. In Forrester's (1961) view, the system dynamics methodology would be most easily employed in small and medium size organizations:

> Last, there is the question of the size of company to which industrial dynamics is most suitable. There is a common first presumption that industrial dynamics is a tool of use primarily to the largest corporations.

As the field has thus been developing, there seems to be little immediate support for this conclusion. In the largest organizations, the functional compartmentalization is apt to be stronger than in smaller companies, making it very much more difficult for any person actually to cut across all activities from research to marketing. It is beginning to appear that the aggressive, rapidly growing, medium and small size organizations may be the places where the methods discussed in this book will have their first important impact. Such organizations are often more flexible. They may be more responsive to the wishes of the company officers, so that if the officers want to explore a new management tool, the organization will indeed do so. In the newer companies, the management is often younger, expects to hold office longer, and takes a longer-range view of developing company strength than in more mature companies. The smaller organizations may be more fluid, so that the rigidities of functional sub-divisions are not so much of a handicap. The costs of management systems research are low enough, so that they present no great difficulty in an organization as small as one million dollars per year of business.[1]

If policy change is a more complex and inertial process in a larger organization, the complexity and slowness of change is even more marked in the area of public policy. In comparison with the typical case of a single organization there is first of all a difference in size. A large number of people are involved in or affected by a public policy issue. Most problems range across several institutional boundaries. The information relevant to analysis of a given policy is more diverse, and consequently requires a wider expertise. Secondly, public policy is formed in a different way than corporate policy (single organizations other than companies will normally represent an intermediate case). The main objective of a company is clear: a good economic result. Whether the result needs to be sustainable, or whether the decision is to make one choice over the other between the long and the short term or between one product line and another may be an open question, but this does not change the largely unidimensional objective. A company is usually hierarchically organized with clearly defined responsibilities on each level. A fairly small group has the power to change policy as long as the major economic objectives are met. On the public policy scene many interests and objectives meet in a process of give and take. Public policy goals are many-dimensional, and are continuously discussed and redefined. Public policy is formed through bargaining between a large number of interest groups, each with many members. The policy-makers, if they can be identified, are dependent upon support of an electorate or special interest group. Significant policy change is preceeded by public discussion.

The described difference between corporate and public policy making do not apply in all cases, but appear as a clear tendency. Frequent complaints that decisions are made over people's heads suggest that there are variations and that public policy is not always formed in a participatory manner. Likewise, there are signs that corporate policy making in several countries is moving in the direction of public policy making. The concentration of business activity to fewer and bigger units increases the influence that an individual company exerts on people's lives. This, and other tendencies, seems to make it increasingly necessary for the companies to supplement the purely economic objectives with objectives in the area of work environment, pollution, job security, and so forth. There also seems to be a gradual erosion of the hierarchical decision structure in favor of more democratic forms of organization.

We can begin to identify the differences between the corporate setting and the public policy scene in terms of the environment for policy analysis. An attempt is made in table 14.1.

In the small and medium size companies the analyst can work closely with a few managers who simultaneously are the main information sources, participants in the modeling process, and the users of the study results.

The problems of integrating information gathering, modeling, and implementation are accentuated when we move into the area of public policy. First, the gathering, processing, and interpretation of information will be so extensive a task that it becomes a research activity in its own right. A particular skill is needed to connect empirical research on the micro- and macro-level. The analyst must be able to converse with a wide range of specialists to obtain the information he needs from historical data categorized according to terminology and theories specific to narrow disciplines. Secondly, only a small number of people can be active in the modeling process. They will have to serve as representatives of the very large number of people implicated by any given policy issue. This "reference group" can ensure that relevant questions are studied, that qualitative information is included in the analysis, and that some users become sufficiently familiar with the research effort so as to be able to make an informed assessment of its quality and finally provide a (narrow) channel for implementation.

But insights from a policy study must be spread to many more people than those who participate in the modeling process, if the study is to have

Table 14.1
Typical differences in the policy analysis environment between corporate setting and the public policy scene

Aspects of analysis	Corporate setting	Public policy scene
1. Owners of problem	Few; closely tied up with problem	Large number; each identifies only partly with problem
2. Objectives of concerned social institutions	Few; clearly defined	Many; diffuse
3. Sources of information	Interviews with employees; corporate records	Close to infinite number of decision-makers and area specialists; historical records; research results
4. Depth and scope of analysis	Limited to the problems of a specific company assuming the framework of the company as given	Includes several "sectors" of society and in each sector several organizations/companies; long time-horizon forces analysis to deal with fundamental social processes, viewing the current social institutions as subject to change
5. Policy options	Few; clearly defined	Many; in the form of broad strategies
6. Participation of policy-makers in analysis	High-intensity participation feasible	Only limited participation feasible
7. Nature of results from analysis	Specific policy recommendations	Framework for broad policy discussions
8. Users of analysis	A few policy-makers and their staff	Many policy-makers and their constituency
9. Evaluation of analysis	Same as users	Same as users plus research community
10. Implementation of policy change	Involves a limited number of key policy-makers and can be carried through fairly rapidly once they have made up their minds	A slow process; policy-makers have to solicit support form an electorate and/or members of several interest groups; institutions must often be changed.

impact. The analyst can not rely exclusively on person-to-person communication with the users of his research findings. Insights gained during the model building must be translated into ordinary language and presented in an easily intelligible form. This requires clarity of concepts and awareness of the reader's way of thinking. The analyst's ability to communicate insights becomes as important as his ability to arrive at them.

System dynamics modeling can become a useful tool also in the area of public policy. This will, however, require that the special characteristics of the public policy scene are recognized and allowed to influence the way in which policy studies are organized and carried out.

This paper reports experience from the use of system dynamics in the study of policy options for the Scandinavian forestry and forest industry. The Scandinavian forest sector is facing a wood resource constraint, and the public policy issue of interest was how to handle the transition from rapid to slow growth in the sector, in a way that would minimize the accompanying problems for workers, management, local communities, and the national economy. The objective was not so clear at the outset. There was no clearly defined client, but instead several diffuse client groups. By reporting the successes and failures of the research methodology chosen in the study, this paper contributes to the discussions around the use of system dynamics in public policy analysis.

14.2 History of the Project

The stated objective of the study was to identify and clarify problems that the Scandinavian forest sector might encounter during the next thirty years or so. Research councils in Sweden and Norway financed the project for the purpose of acquiring experience with the usefulness of system dynamics in public policy analysis. The project lasted for two years and was carried out by a research team of three persons, all trained in system dynamics and one having some acquaintance with the forest sector. A steering committee for financial control and methodological advice was set up. Close contacts with Scandinavian forest economists were established before the project began.

The first two months of the project were devoted to formulating a research plan. Contacts were established with representatives of various interests and researchers in forestry and the forest industry.

The original research plan envisioned a project that would be carried through in three phases. During the first half year the research team would define the problem focus in close cooperation with a selected group of decision-makers. The research team expected to develop a crude simulation model that would serve to pinpoint the chosen problem. The model would illustrate the kind of problems expected from adherence to current policies. Examination of this first very tentative model was expected to give rise to questions about the validity of certain model assumptions, and make possible a selection of those topics most in need of detailed research. A whole year was allocated for detailed research on the chosen topics. During the last half year, the plan was to use sub-study results to upgrade the initial model, and then to analyse alternative policies in cooperation with decision-makers. A main goal was to present study insights from the project *without* reference to formal models, so they would be intelligible to a wide audience.

The actual progress of the project differed from the plan in two important ways. First, development of the initial simulation model proved to be more time consuming than expected. Second, it was difficult to assign meaningful tasks to researchers outside the team, partly due to the slow development of the initial simulation model. As a result the sub-study phase was reduced and most of the detailed research was carried out by in-house team members consulting outside researchers.

After the initial planning period the research staff approached several high-level decision-makers representing labor unions, corporate management, forest owners, research institutions, and government authorities. Reference groups of approximately eight people each were formed. A series of ten half-day meetings over a five-month period focused on current problems, concerns for the future, and interpretations of past development in forestry and the forest industry.

Discussions in the reference groups covered a lot of ground and were experienced as fruitful. Though very loose in the beginning, the meetings became more structured as time went by. The only modeling tool employed was the causal loop diagrams. The end-product of this series of meetings was a problem definition judged relevant by the reference groups: there exists a need to limit forest fellings in Scandinavia in the near future. Such stabilization of the industrial use of wood will, however, when combined with increasing labor productivity lead to falling forest sector employment. A tempting short-term solution will be to allow fellings to exceed their sustainable level. The long-term effects are undesirable: a forced reduction in production volume and faster reduction in employment than if expansion of the industry beyond the sustainable level had been avoided.

Having defined the problem, the team began working on a simulation model. Four months were needed to arrive at a first running model, much more than expected. The delay was partly due to an early effort to develop a much too general model. This was particularly true of that part of the model that represented the forest industry. Instead of focusing the modeling on those changes in the forest industry that were directly connected to the industrial use of wood, something close to a general production function for an industry was attempted. The modeling progressed rapidly once the research team concentrated on the primary effects of a limited wood supply.

The resulting model covered most aspects of forest sector development that came to be regarded as central issues during the remainder of the project. It proved, however, difficult to derive from the model specific questions that would be suited for analysis in other research groups. The model concepts were too fluid to serve as guidance for detailed empirical research by outside specialists. Modeling was therefore continued, and the emphasis on sub-studies was reduced.

The first model version had several weak points. The simulation runs were too sensitive to variations in parameter values. Scenarios generated by the model were unrealistic in important ways. The forest industry would, for example, in some runs enter into a vicious circle of low profitability reducing investments, insufficient investments further lowering profitability, and so on, until the industry was practically wiped out. In the real world this process could be expected to be strongly buffered by corrective actions in the environment of the forest industry through changes in exchange rates, wages, prices, or government subsidies. In the model the buffers were weak and limited to changes within the forest industry. The team tried to improve the model by adding new structure and making the old structure more detailed. As a result the model grew in size. It became harder to work with and less transparent. After a vacation twelve months into the project, an attempt was made to sort out the basic ideas represented in the model. A new start was made and the central concepts and mechanisms were assembled into a simpler and clearer model.

This second model also had a limited life span. After some months the team once again started on a new model. This process repeated itself throughout the project. A number of models were developed, each with a total life of two or three months. The step from one model to the next always seemed drastic. In retrospect, there appears, however, to have been a high degree of continuity. Each model shift represented a step forward in terms of conceptual clarity.

Throughout this extended modeling process, the team felt a need to sum up their assumptions and insights verbally, both in order to gain a clearer sense of direction and to get a response from outsiders. Writing is time consuming, however, and therefore required some extra motivation such as requests for conference presentations. The pressure to summarize our current thoughts on paper helped us to keep a tighter focus in the modeling work.

A wide variety of outside sources were consulted throughout the project in order to obtain qualitiative and quantitative information about the historical and current situations in the forest sector. During the first half of the project the empirical work was rather fragmented and directed toward answering specific questions raised by the modeling. Later on in the project the team made systematic efforts to acquire a clear picture of how the Scandinavian forest industry historically had adjusted its production capacity to the available forest resources.

Contacts were maintained with the reference groups throughout the project. Simulation runs generated by the model were discussed during some ten meetings at a later stage. Presentation of scenarios of possible future development patterns turned out to be a very effective way to stimulate the reference groups to put forward and reconsider theories about the past and the future.

The study produced results of two different kinds. First, the members of the reference groups found that the group discussions increased their ability to see their own situation in a larger context and put them in a better position to evaluate the long term effects of various policies. Second, the team produced a number of documents of value for future policy discussions. The most important is a popular book discussing the transition from rapid to slow growth of the Scandinavian forest sector and the likely effects of various policies intended to alleviate transition problems.[2] Documentation of the last version of the model and a large number of working papers are also available, primarily for other researchers.

The coming years will show to what extent the research effort has had any influence. Several research institutes are carrying the research work further. Forest economists in Finland, Norway, and Sweden have decided to apply the model to smaller regions and study the transition from ample to scarce wood supply in more detail at the local level. Corporations, labor unions, government agencies, and other organizations have shown interest in strategy consultations with the research team. Such contacts could offer opportunities to relate the long-term perspective of the project to concrete policy making.

14.3 The Modeling Process

Choice of Problem Focus

A policy study must focus on a limited number of problems in order to yield nontrivial results. The choice of problem will naturally be a major determinant of the relevance of the whole study.

When the forest study began, the research team had not yet chosen a specific problem on which to focus their research. They had a list of emerging problems in the Scanadinavian forestry and the forest industry; they realized that most of them were interrelated but had only vague ideas about how.

The team used the first half year to choose a problem. They discussed in meetings with the reference groups what might become the most important problems during the coming thirty years, and how those problems could be dealt with. The team had to sort out temporary changes from persistent trends, attempt to explain the forces behind the trends, and then hypothesize about what kind of future would emerge in the absence of any drastic changes in policy. At subsequent meetings they would present theories and receive criticism or support.

In retrospect, the decision to allocate the first half year to exploratory problem definition appears wise. It gave the necessary direction for the rest of the project. Such direction is particularly needed in a modeling project. The model builder much too easily loses sight of the objectives of his work, and begins to develop a general purpose model that aspires to answer all questions but in the end yields disappointingly few insights.

The problem focus was cast in the form of a "dynamic hypothesis."[3] This implied hypothesizing simultaneously the future mode of behavior of key variables (the reference mode) and the driving forces (the basic mechanism) behind this behavior. The idea of a dynamic hypothesis was useful. In three respects its employment could, however, have been improved. First, more concrete variables should have been selected. "Structure of industry" should, for example, have been specified to "number of plants," "sawmill production as a fraction of total forest industry production," or the like. Second, the dynamic hypothesis should have been verbalized in detail. The causal diagram employed in the project left too many questions unresolved. Third, more attention should have been given to the quantitiative aspects of the reference mode. Through simple hand calculations it would have been possible to make order-or-magnitude estimates of the likely development of important variables.

Scope of Model

During the following year and a half of the project most of the work centered around the development of a system dynamics model of the transition from rapid to slow growth.

The modeling work to some extent distracted the research team from seeing the problem in a balanced perspective. Certain processes, for instance, the geographical concentration of the activities in the forest sector, can be adequately explained only in terms of overall national social and economic development. The only effective means for influencing the rate of concentration are *national* policies. Modeling of the forest sector itself absorbed so much energy that analysis of the interface between the forest sector and the rest of the economy was neglected.

After more than half a year of modeling, the team recognized the difficulty of analyzing the forest sector in isolation. They began to consider feedbacks through exchange rates, government subsidies, and local resistance to close-down of industrial plants, but their analysis was so late in the project that these issues were given less than due attention. Further, forest sector problems are not commonly discussed in such a wide perspective. To make connections that are not usually made, the analyst must have experience and confidence.

Conceptualization and Empirical Research

Many models were developed in the course of the project. The first were general and abstract, and were never carried to the stage of simulation experiments. In the first running model, the team had reduced the level of generality and based the conceptualization on concrete concepts like "unemployment" rather than nebulous terms like "transition problems." High levels of generality make it difficult to parametrize the model and for other people to assess its validity.

Many of the model-building problems arose from attempts to conceptualize on too scanty empirical ground. Lacking empirical information, the team often unconsciously moved their conceptualization to a more abstract level where the particular detailed characteristics of the analyzed topic became less significant. They could then proceed with analysis, but reached few insights of value for policy making. Since the concepts lacked specific empirical content, the analysis had too much the character of pure mathematics of dynamic systems. Sooner or later, the team intended to acquire the information necessary to develop more concrete concepts. In the meantime, however, they wasted considerable efforts on modeling when they should have been conducting field work to find out what actually takes place in the real world.

According to the initial project plan empirical research on specific topics should be carried out by outside researchers. As this idea was gradually de-emphasized, the in-house staff should have intensified its own empirical research to provide a firmer basis for the modeling.[4]

Important insights can be gained from comparing development on the micro- and the macro-level (relative terms, of course). In the forest study the research team had a tendency to remain on the macro-level and deal exclusively with aggregates. For example, they often confused the dynamics of a single production plant with that of the whole industry. They did not complement aggregate statistics with enough plant studies. There was a similar situation in analyzing forest growth, although detailed study of the growth dynamics of an individual stand in the beginning of the project did provide some basis for aggregate analysis.

Verbal Analysis as the Basis of Model Building

When using a formal modeling technique, such as system dynamics, a modeler is tempted to substitute modeling for verbal analysis. This tendency generates a lot of unproductive modeling. Fruitful concepts are created through verbal analysis. Representation in a formal model can help to identify inconsistencies, incompleteness, or ambiguities but can not in itself create fruitful concepts. Conceptual analysis should be carried out in purely verbal form. Formal analysis should be used to complement and accelerate verbal analysis.

In the forest study the team did not spend enough time on verbal analysis. Too often, they found themselves shuffling around abstract concepts with obscure ties with reality. Once a concept has been positioned in a formal structure, it becomes frozen, and its further use becomes very inflexible. This inflexibility may be an advantage once fruitful concepts have been developed and the modeler wants to preserve their meaning to facilitate analysis and communications. But not until then!

Transparency of Model

A conceptualization is more valuable the simpler it is as long as it remains true to the essence of the problem under investigation. The initial decision to view a nontechnical report rather than a simulation model as the major product output from the project motivated the research team to keep the model simple and transparent. Simplicity was achieved in several ways. First, the model was built with a firm view toward the prime objective: to test the dynamic hypothesis. This allowed the exclusion of various irrelevant or less important relationships. Second, the model was restricted to processes with time constants relevant to the dynamics of the transition from ample to scarce wood supply. Fluctuations in capacity utilization and wood prices connected with the regular business cycles were, for example, considered to be too rapid processes and were smoothed over. Third, variables with the same "dynamic function" were aggregated into one variable.

14.4 Reference Groups

Reference groups consisting of decision-makers in the forest sector were an important feature of the forest study. The reference groups served as a kind of mini-universe of the part of the real world under study. The groups were assembled in order to enrich the empirical basis of the analysis and to open up channels for communicating insights. The main reasons for bringing people together in groups rather than consulting them on an individual basis were

1. interaction in groups would bring out problems of communication, as well as actual and potential conflicts of interest,
2. group interaction would also make apparent consensus or disagreement about various points of view,
3. institutionalized contacts with respected decision-makers would add to the project's status and credibility.

If a larger number of the group members had been active outside the forest sector, the research would probably have concentrated more on the interface between the forest sector and the rest of society. Such a concentration might have been desirable.

The members of the reference groups experienced their participation in the project very positively. They got a chance to discuss matters with people who were usually their opponents in a setting requiring a minimum of tactical considerations. The meetings gave an opportunity for the group members to consider their own situation in a larger context and with a longer time perspective than usual. The sessions centering on simulation runs were found the most thought-provoking.

The discussions remained often quite general, however. To some extent this tendency reflected the early stages of the conceptualization process. Further, the research team only had a vague impression of the daily work situation of the group members. Early contacts with individual group members, focusing on their current work and special experiences, would have helped make later discussions more concrete and alive.

It was further found difficult to attain continuity in the meetings—both time-wise and content-wise. This was largely due to an excessive emphasis on the model. The strong focus on model development made group contacts unnecessarily irregular. The team was constantly post-poning contacts, hoping that next week or month they would have a much improved simulation model running. Since the discussions in the reference groups were planned to center around model experiments, contacts with the reference groups became difficult to plan.

The meetings where simulation runs were discussed could have been more effective if there had been a higher degree of continuity. After outputs from a given model had been discussed at two or three consecutive meetings, there was a break either in the sense that the model was changed drastically or that the next meeting was scheduled for a much later occasion in order to give enough time for the research team to process new ideas that had come up.

The incessant worry about how to maintain the interest of the reference group members so they would continue to attend meetings made it impossible for the team to have a period of sorely needed uninterrupted work.

Finally, the reference groups had some difficulties in appreciating and using the aggregate and continuous perspective of the research team—for example, in the case of description of policies. In practice policies are seldom formulated explicitly. They often have the character of fairly simple rules of thumb. Policies may be very hard to state explicitly, since they only appear implicitly in individual decisions, each of which is embedded in complex situations where the particular is hard to distinguish from the general. Some decisions, like the close-down of plants, are so rare for the individual decision-maker that each instance appears unique. Consequently, there does not seem to be any common policy behind the individual decisions. In this and similar cases the policies would have stood out more clearly if a cross-sectional analysis of recent events had been complemented by discussions of the history of some individual companies, forest holdings, and the like. A comparison of strategies outlined in policy documents ten or twenty years ago with the development that had actually taken place would probably also have added to an understanding of what factors influence various decisions.

Even if the reference group members supplied a wide range of experiences, in many cases the team had to seek information from external sources. Consultations ranged from telephone calls to full-day meetings.

14.5 Information Sources

The study rested on three sources of information: statistical data, descriptive written material, and oral reports given in a discussion. All sources are important in spite of the difference in precision levels. Statistical data provide quantitative estimates of a few carefully specified phenomena. Written sources bring forth a wide spectrum of qualitative information relevant to a certain issue with a high degree of semantic precision. In a discussion the level of precision is even lower. On the other hand, it leaves open which issues are to be dealt with, making it possible to decide this as the exchange of information proceeds, and allowing for an active learning process. The lesson from the forest study is that all three sources should be used, but the analyst should keep in mind the proper role of each one.

During the first half year of the project the research team had numerous meetings with other research groups on issues of forestry and the forest industry. Study plans were exchanged, and there seemed to be many areas of mututal interest. The initial model building, however, required longer time than expected, and short of a running model, it proved difficult to select a few questions for more thorough analysis by some outside group. During the first year, model concepts were so unprecise and constantly changing that it was found less committing to obtain information in short person-to-person discussions.

To help orient themselves in the industry and in the research community the team had a Norwegian forester working with them during large parts of the project and frequently consulted with a Swedish professor in forest economics. These two persons gave easy access to the "conventional wisdom" of the forestry and the forest industry communities. They also provided valuable advice about where to search for specific information.

Initially, the team studied history through statistics. Successively, increasing attention was given to how people perceived their own situation in the past, what they valued as important, and which policies were actually agreed upon and followed. This information could not be extracted from quantitative data but was available in old government reports and historical monographs. This helped the team to a more nuanced picture of both the past and the present.

It was only much later that the team discovered how useful professional historians are in establishing a general picture of past developments. Historians can give a systematic introduction to the literature and help select the most informative works. And probably more important, the historians can recount images of the effects past policies had on forestry and the forest industry.

The staff derived little benefit from the research projects with which they had established contacts earlier on. This was partly due to a low appreciation of empirical information at the beginning of the project, in turn connected with a commitment to the system dynamics dogma teaching reverence for structures as opposed to parameters, which was wrongly taken to mean that empirical studies are unimportant. Collection of high precision data is, indeed, a waste of effort in the early stages of conceptualization. Choice of fruitful concepts requires, however, enough empirical knowledge so that crucial factors can be distinguished from those of marginal importance.

A researcher's total experience and understanding goes far beyond what is recorded in reports of specific research projects. As a consequence, the team found personal contacts with other researchers a necessary complement to, and often more rewarding than, the study of written material. In writing a report, the author would have addressed a specific issue that might have little in common with the team's problem. Research reports were often full of loose ends the team wanted to pursue further. For this, direct conversation with the report writers proved indispensable. We gradually learned that it is difficult to establish communication beginning on an abstract level. A discussion that started out around some very specific matter, and then widened in scope, usually provided more generally applicable insights.

14.6 Recommendations

Based on our experiences with the forest study, we can suggest one procedure for the use of system dynamics in public policy analysis in the form of an ideal research procedure for a modeling project. It is assumed that the available research time and manpower resources as well as the scope and complexity of the policy issues are similar to those in the forest sector study. To simplify the presentation, we distinguish between four broad, and in practice overlapping, activities: model building, empirical research, interaction with the reference group, and report writing. Figure 14.1 shows the effort that would go into each of the four activities during different phases of the project. The relative importance of the four activities varies with time in the ideal project plan. Accordingly, five different phases of the project can be identified: problem definition, development of initial model, empirical research, model improvement, and report writing. These are all subsequent to a low intensity pilot study.

Problem Definition

The purpose of the problem definition phase is to work out a focus for the research. It is useful to define the problem in the form of a dynamic

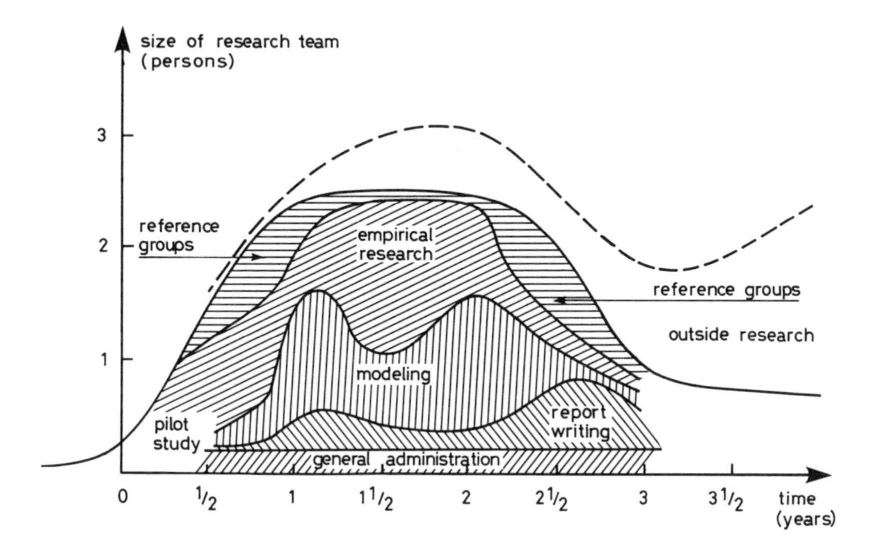

Figure 14.1
Revised study plan

hypothesis, that is, to specify the expected development pattern of central variables and give a causal explanation of why this development is likely.

A series of discussions in a reference group consisting of people with experience from the problem area will yield a comprehensive picture of concerns about the future, interpretations of the causes of past and current problems, and policy proposals for the future. Individual contacts with decisions-makers and researchers and the study of written material can be used to complement the information provided by the reference group. People and documents that can present insightful interpretations of the historical development will be particularly helpful at this stage.

When fundamental problems and conflicts can be distinguished from surface symptoms, it is possible to define an appropriate time-horizon for the study and, tentatively, define a set of key variables. Statistics should be processed to obtain time-series for the variables covering a period corresponding to the time-horizon of the study.

Very simple models may be built to illustrate the dynamic effects of feedbacks and delays and to introduce the members of the reference group and others to the basic ideas of modeling.

The output of the problem definition phase should be a written statement of the problem focus, supported by an account of relevant aspects of historical development in the problem area and rough projections of key variables into the future. It is important that the reference group finds the problem relevant and in need of further analysis.

Development of an Initial Model

When a problem focus has been arrived at, a system dynamics model illustrating the problem should be developed within three or four months. If the problem statement is sufficiently concrete, the modeling will be mainly a matter of formalization of a verbal theory. The initial model should be developed quickly, and the level of detail and conceptual clarity adjusted to this requirement. The success of this phase will depend on whether the team can keep the model small and well focused. To avoid falling in the trap of building an ever larger model, it may be useful to think of the modeling as nothing but a means of expediting a set of clearly defined calculations too complex to be carried out by hand.

A modeler who has clear expectations about the behavior of a model will readily discover, and take seriously, surprising effects exhibited by a simulation run. The strange behavior—explosions, collapses or violent oscillations—often generated by early models can be a sign that important control mechanisms operating in the real world have by-passed the modeler.

The empirical basis for the initial model will naturally be weak. The process of building and running a model will raise questions of the relative importance of various phenomena, and in this way guide the ensuing empirical research.

Although the model is intended primarily for project planning purposes, it should be documented.

Empirical Research

A system dynamics model will typically represent the interdependence between a number of "populations." In a study of medical care the populations may, for example, be hospital buildings, equipment, doctors, nurses, patients, and potential patients. The represented characteristics of each population depend on the particular problem under study. Usually, the size of the population in some sense is important. In addition, there may be one or more measures of "intensity" such as age of buildings, productivity of equipment, skill of medical personnel, or resistance to disease among the population. Changes in one population is tied to the population itself and the other populations through "policies," some of which are given by nature and others which are instituted by humans and subject to change.

In a system dynamics model it is easy to distinguish between the flows and accumulations for each population (the flow structure) and the policy structure that governs the flows. In our revised study plan the initial model will contain a rough outline of a flow structure and policy structure based upon the analyst's general understanding of the situation and the most readily available opinions and material concerning the problem area.

The purpose of the empirical research following the initial modeling is to give the analyst a richer and more accurate picture of those aspects of the real world that contribute to the dynamic behavior of the initial model. Parallel study of the lifecycle of individuals and the dynamics of aggregates of individuals (populations) will indicate useful disaggregations of the populations. Realism in the representation of policies requires that the analyst knows what information is available to the various decision-makers and how they use it. Also here micro and macro analysis should be combined.

Most of the empircial research will be carried out by the research team. This is necessary because the organizing concepts are fluid and may be frequently changed, so that the research strategy may have to be altered as the empirical research yields new insights about the system structure and behavior. If possible, the team should arrange to work closely with some outside researchers who have contacts with specialists and knowledge about the organization of information in the area.

The empirical research will include such diverse activities as interviews with decision-makers, study of historical records, discussions with outside researches, and compilation of statistics. Modeling may be used in connection with the empirical research, primarily as an accounting device to aid in historical analysis of material flows, population dynamics, and so on.

Model Improvement

The ultimate objective of this phase is to improve the mental model of the members of the research team and of the members of the reference group. The means for achieving this objective is experimentation with an upgraded version of the initial model that incorporates the findings from the empirical research.

The first three to four months go into building an improved model which can generate scenarios that will serve as a basis for later discussions in the reference group. If the empirical research has been properly carried out, the staff will approach the modeling with concrete images of the lifelines of typical individuals in the populations to be modeled, and furthermore have a failrly accurate idea about the extent of the variations around those typical cases. Particular questions may have to be researched as they come up, but on the whole the model building should be a surveyable and predictable task for which the level of ambition can be adjusted to the available time.

Clarity of model concepts is important if the discussions in the reference group are to be effective. The model should be kept as simple as possible and still represent the important dynamics. A model with many "decorative variables" will confuse issues more than clarify them. If the modeling is focused on the most basic mechanisms, the resulting model will tend to be robust, that is, it will behave in a reasonable fashion even when extreme changes in policies or parameters are made.

When a robust model is running and well understood by the team, it will be useful to start the second series of meetings with the reference group. The purpose of the discussions in the group is to increase the group members' understanding of the problems they are facing and develop a basis for the popular project report.

The model is used to generate consistent and reasonably realistic scenarios. Simulation runs will show the combined result of growth and decay processes, physical and social limits, information delays, and nonlinear relations between system variables. The purpose of the model experiments is to help provide insights into how various policies, due to their position in a complex structure, will tend to alleviate or amplify problems. The purpose is not to produce predictions.

The members of the reference group will compare the behavior of the model with their intuitive ideas about the functioning of the real world. The comparison will at least in the beginning be very impressionistic. It will typically concern the relative phasing in time of various phenomena, the amplitude of oscillations or overshoots, and the rate at which adjustments are made. When the mental models of the group members yield different results than the model experiments, the team should be able to show why the simulation model behaves the way it does, making it possible for the group members to criticize or accept the assumptions made in the model. In the latter case the group members will have changed their mental models. One simulation run can often be enough to carry productive discussions through a half-day meeting.

Between reference group meetings the team will work to clarify issues that were raised at the previous meeting. This may take the form of empirical research, model improvement, or authoring short memoranda. The criterion for judging whether a specific model change should be introduced or not is the extent to which it will enlighten reference group discussions.

When the group members have become familiar with the model, they will be able to suggest realistic policies to alleviate undesirable behavior shown in model runs. Much can be learned from testing the effects of these policies in the model system.

Report Writing

Along with results from the empirical research the discussions in the reference group form the basis for the major report of the project. This report should address a wide audience and be written in a nontechnical manner. It will typically contain an interpretation of past development and present alternative scenarios of the future. It should attempt to explain the connection between emerging problems and the structure underlying them, and show how various proposed policies might alleviate or amplify the problems.

Parallel to writing the report, the model should be finished up and documented with an explanation of its rationale equation by equation. The documented model along with working papers from the project will be of value primarily to other researchers.

Notes

1. J. W. Forrester, 1961, p. 365.

2. J. Randers, L. Stenberg, and K. Kalgraf, 1978, *Skognærigngen i overgangsalderen* (Oslo: J. W. Cappelens Forlag).

3. See J. Randers, 1973, p. 54.

4. Compare J. Randers, 1973, p. 244: "However, when encountering modeling problems, one is easily trapped into believing that the obstacle is the limited capability of the modeling tools to represent reality. Unending, futile attempts at formulating some part of the model is symptomatic of a lack of knowledge of the real system."

References

Forrester, Jay W. 1961. *Industrial Dynamics.* Cambridge, Mass.: MIT Press.

Randers, Jørgen. 1973. *Conceptualizing Dynamic Models of Social Systems: Lessons from a Study of Social Change.* Ph.D. dissertation. Alfred P. Sloan School of Management, MIT, Cambridge, Mass.

Randers, Jørgen, Lennart Stenberg, and Kjell Kalgraf. 1978. *Skognæringen i overgangsalderen.* Oslo: J. W. Cappelens Forlag.

Index

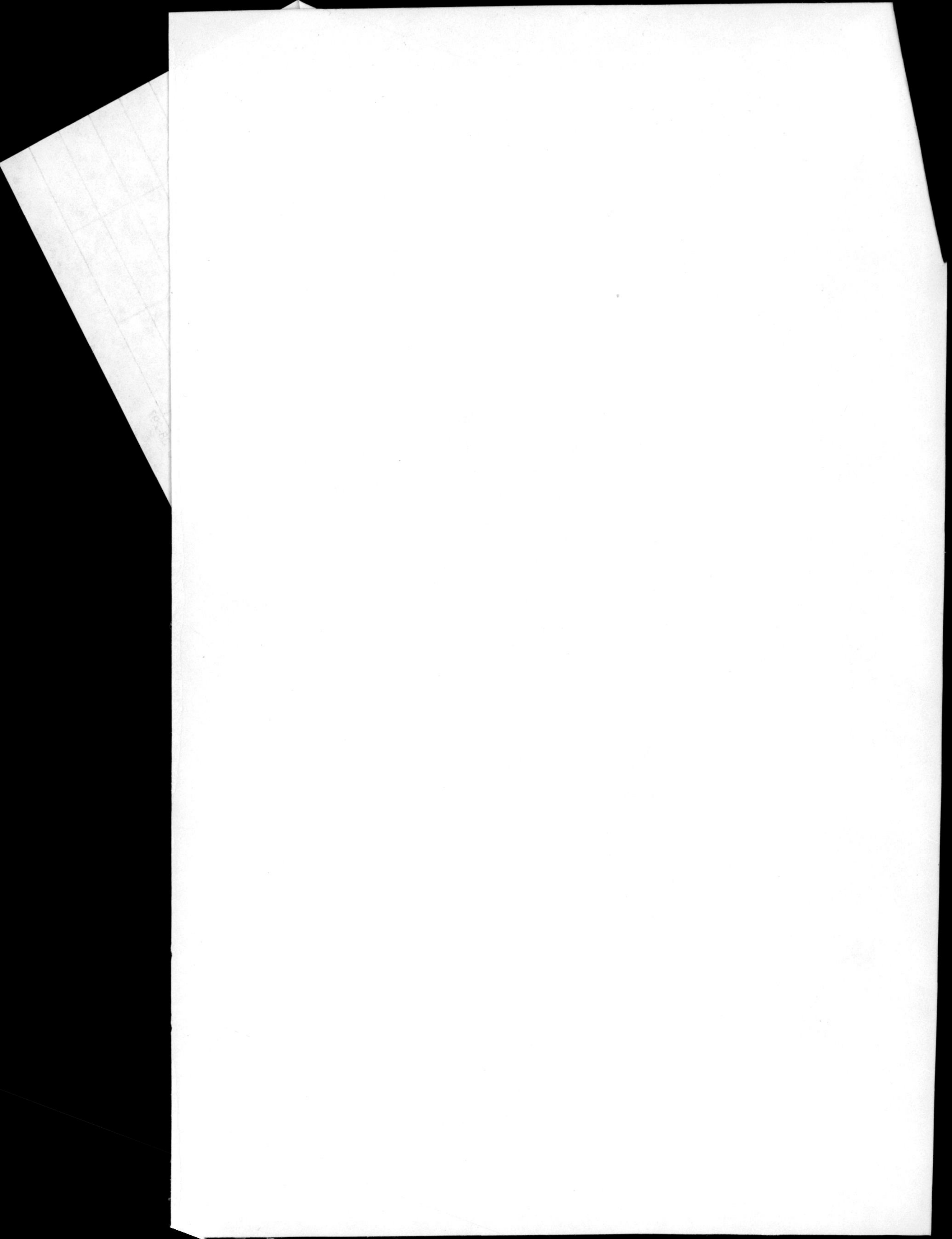

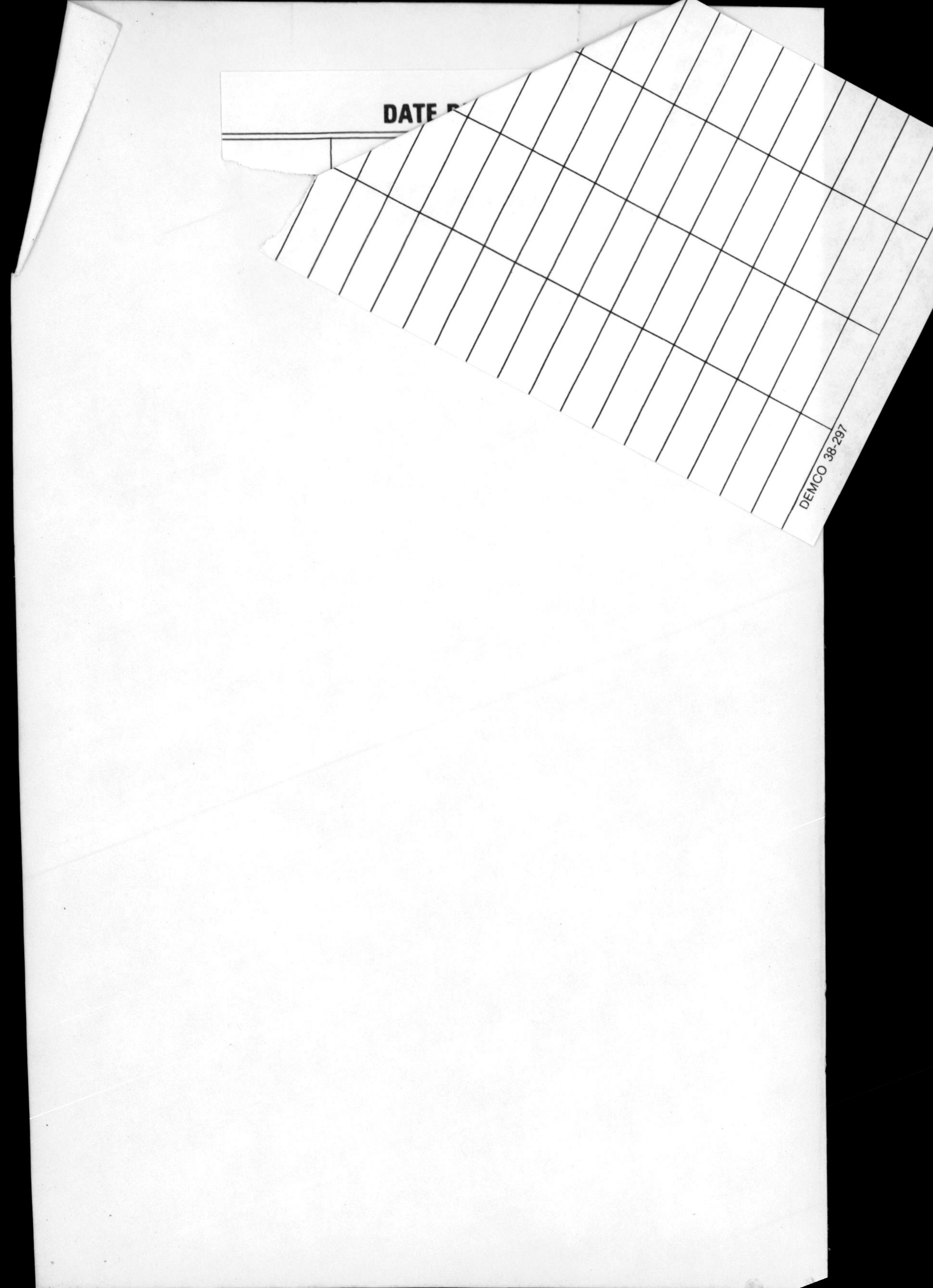
DATE D
DEMCO 38-297